H.-D. Höltje, W. Sippl, D. Rog
and G. Folkers
Molecular Modeling
Basic Principles and Applications
Second Edition

LIBRA

This book
and may

Related Titels from Wiley-VCH

W. Koch, M. Holthausen

A Chemist's Guide to Density Functional Theory
(2nd edition)

2001

ISBN 3-527-30372-3

C. W. Sensen

Essentials of Genomics and Bioinformatics

2002

ISBN 3-527-30541-6

N. Sewald, H.-D. Jakubke

Peptides: Chemistry and Biology

2002

ISBN 3-527-30405-3

J. Gasteiger, T. Engel

Chemoinformatics – A Textbook

2003

ISBN 3-527-30681-1

Hans-Dieter Höltje, Wolfgang Sippl,
Didier Rognan, and Gerd Folkers

in collaboration with Robin Ghosh and Pavel Pospisil

Molecular Modeling

Basic Principles and Applications
Second Edition

WILEY-VCH

WILEY-VCH GmbH & Co. KGaA

Prof. Dr. Hans-Dieter Höltje
Institute of Pharmaceutical Chemistry
Heinrich-Heine-Universität
Universitätsstr. 1
D-40225 Düsseldorf
Germany

Prof. Dr. Wolfgang Sippl
Faculty of Pharmacy
Martin-Luther-Universität Halle-Wittenberg
D-06099 Halle/Saale
Germany

Prof. Dr. Didier Rognan
Faculty of Pharmacy
Université Louis Pasteur Strasbourg I
74 route du Rhin
F-67401 Illkirch-Graffenstaden
France

Prof. Dr. Gerd Folkers
Institute of Pharmaceutical Sciences
ETH Zürich
Winterthurerstr. 190
CH-8047 Zürich
Switzerland

■ This book was carefully produced. Nevertheless, authors and publisher do not warrant the information contained therein to be free of errors. Readers are advised to keep in mind that statements, data, illustrations, procedural details or other items may inadvertently be inaccurate.

1st edition 1996[1]
2nd edition 2003

1) published as Vol. 5 of the series *Methods and Principles in Medicinal Chemistry* (series editors R. Mannhold, H. Kubinyi, H. Timmerman)

Library of Congress Card No. applied for.
British Library Cataloguing-in-Publication Data:
A catalogue record for this book is available from the British Library

Die Deutsche Bibliothek –
CIP Cataloguing-in-Publication-Data
Bibliographic information published by
Die Deutsche Bibliothek
Die Deutsche Bibliothek lists this publication in the Deutsche Nationalbibliografie; detailed bibliographic data is available in the Internet at http://dnb.ddb.de

© 2003 WILEY-VCH Verlag GmbH & Co. KGaA, Weinheim

Composition Kühn & Weyh, Satz und Medien,
             Freiburg
Printing and Druckhaus Darmstadt GmbH,
Bookbinding Darmstadt

Printed in the Federal Republic of Germany

ISBN 3-527-30589-0

Inhalt

Preface to the second edition

"Science may be described as the art of systematic oversimplification."
Sir Karl R. Popper

When in 1995 we sketched the first draft of the first edition of this book, we discussed many objects to be dealt with, that at those recent times were a kind of "secret knowledge", spread among beginners in the art of molecular modeling by word-of-mouth recommendation. In less than a decade, the scenario has changed dramatically. It seems there is simply no research group in molecular science, be it in academia or industry, which is not using 3D color representations of their molecules. Moore's law has bestowed us with high performance notebooks able to visualize MD trajectories in a classroom presentation. Hence, most students acquired knowledge about and have become familiar with molecular modeling. Lecture notes and work reports are filling up with colorful molecules and, in the era of paperless offices and wireless communication, so are the hard disks of laptops. What has not changed is the fact and our shared conviction that it might be good to know where all these illustrative molecules come from and what is behind the colorful surface.

While the first edition was meant for PhD students, being the modeling beginners at those times, this edition is clearly addressed to undergraduate students. The goals have not changed, nor has the intention. The contents of the book, however, have been updated, new developments have been added without making it a tome and the price has been coming down to meet the needs of a student audience.

We express our gratitude to our co-authors and collaborators as well as to the publisher, Wiley-VCH, for the trustful relationship and the flexibility in developing this project.

May 2003
Düsseldorf
Zürich

Hans-Dieter Höltje
Gerd Folkers

Preface to the first edition

"A Model must be wrong, in some respects, else it would be the thing itself. The trick is to see where it is right."

Henry A. Bent

We humans receive our data through the senses of vision, touch, smell, hearing and taste. Therefore, when we have to understand things that happen on the submicroscopic scale, we have to devise a way of simulating this activity. The most immediate and accessible way to represent the world that is unobservable is to make a model that is on our scale and that uses familiar forms.

Many physical and chemical properties and behaviors of molecules can be predicted and understood only if the molecular and electronic structures of these species are conceived and manipulated in three-dimensional (3D) models. As a natural follow up nowadays the computer is used as a standard tool for generating molecular models in many research areas.

The historical process of developing concepts leading to molecular modeling started with the quantum chemical description of molecules. This approach yields excellent results on the ab initio level. But the size of the molecular systems which can be handled is still rather limited. It is therefore that the introduction of molecular modeling as a routine tool owes its beginning to the development of molecular mechanics some 25 years ago together with the appearance of new technologies in computer graphics.

The goal of this book is to show how theoretical calculations and 3D visualization and manipulation can be used not simply to look at molecules and take pretty pictures of them, but actually to be able to gain new ideas and reliable working hypotheses for molecular interactions such as drug action.

It is our intention to reach this goal by giving examples from our own research fields more than reporting literature's success stories. This is because stepwise procedures avoiding pitfalls and overinterpretation can at best be demonstrated by data from our own laboratory notebooks.

Most of the contents will therefore reflect our own ideas and personal experiences, but nevertheless represent, what we believe to be an independent view of molecular modeling.

We gratefully acknowledge the technical assistance of Matthias Worch, Frank Alber and Oliver Kuonen. Finally we wish to express our sincere gratitude to Heide Westhusen for her excellent secretarial and organizational help.

Spring 1996
Berlin   *Hans-Dieter Höltje*
Zürich   *Gerd Folkers*

1
Introduction

"Dear Venus that beneath the gliding stars ..." Lukrez (Titus Lucretius Carus, 55 B.C.) starts his most famous poem *De Rerum Natura* with the wish to the God-dess of love to reconcile the wargod Mars, which in this time when the Roman Empire starts to pass over its zenith, ruled the world.

Explanation is the vision of Lukrez. His aim is in odd opposition to his introductory wish to the goddess of love: the liberation of people from his fear of God, from the dark power of unbelievable nature.

The explanation of mechanism from the common is the measure with which Lukrez will take away the fear from the ancient people, the fear of the gods and their priests, the fear of the want of nature and the power of the stars.

Lightning, fire and light, wine and olive oil have been perhaps the simple things of daily experience, which people needed, which people was afraid of, whom has been dear to him:

"... again, light passes through the horn
of the lantern's side, while rain is dashed away.
And why? – unless those bodies of light should be
finer than those of water's genial showers.
We see how quickly through a colander
the wines will flow; how, on the other hand
the sluggish olive oil delays: no doubt
because 'tis wrought of elements more large,
or else more crook'd and intertangled ..."

The atom theory of Demokrit leads Lukrez to the description of the quality of light, water and wine. For this derivation of structure–quality relationships he uses models. The fundamental building stones of Lukretian models look a little like our atoms, called *primodials* by Lukrez, elementary individuals, which were not cleavable anymore. Those elementary building stones could associate. Lukrez even presupposes recognition and interaction. He provides his building stones with mechanic tools that guarantee recognition and interaction. The most important of these conceptual tools are the complementary structure (sic!) and the barked hook. With these primordials Lukrez built his world.

How well the modeling fits is shown in his explanation of the fluidity of wine and oil. A comparison of the space-filling models of the fatty acid and water molecules amazes, because of its similarity with the 2000-years old image of Lukrez.

1.1
Modern History of Molecular Modeling

The roots from which the methods of modern molecular modeling have developed, lie at the beginning of our century, the first successful representations of molecular structures being closely linked to the rapid developments in nuclear physics.

Crystallography was the decisive line of development of molecular modeling. Knowledge of the complexity of crystal structures increased very rapidly but their solution still required huge arithmetic expense to produce only an inadequate two-dimensional (2D) paper representation. The use of molecular kits was the only possible way of obtaining a 3D impression of crystal structure.

The Dreiding Models became famous because they contained all the knowledge of structure chemistry at the time. Prefabricated modular elements, for example different nitrogen atoms with the correct number of bonds and angles corresponding to their hybridization state, or aromatic moieties, made it possible to build up very exact 3D models of the crystal structures, thus allowing molecular modeling. Dimensions were translated linearly from the Ångstrom area. Steric hindrances of substituents, hydrogen bond interactions, etc. were quite well represented by the models. A similar quality of modeling, albeit less accurate—but space filling—was provided by Stuart–Briegleb or CPK models. Watson and Crick described their fumbling with such molecular kits and self-constructed building parts, first to model base pairing and eventually, to outline the DNA helix.

Molecular modeling is not a computer science a priori, but does the computer provide an additional dimension in molecular modeling/molecular design? Indeed, development of the computer occured synergetically, as faster and faster processors repeated the necessary computational steps in shorter and shorter times so that proteins containing thousands of atoms can easily be handled today. However, the molecular graphics technology looked for a further quantum leap bound to the same fast processors. For the first time, in the 1970s the pseudo 3D description of a molecule, color-coded and rotatable, was possible on the computer screen. "Virtual Dreiding models" had been created. Without computer technology the flood of data emerging from a complex structure such as a protein would have exceeded the saturation limits of human efficiency. Proteins would not have been measurable with methods such as X-ray structure analysis and nuclear magnetic resonance without the corresponding computer technology. Indeed, it is computer technology that has made these methods what they are today.

There is however a second factor, without which today's computer-assisted molecular design would be unthinkable. Since the 1930s, nuclear physics has required not only analytical but also systematic thought, a component that was vital in construction of the atomic bomb. Consequently, mathematical modeling techniques were employed for the computation of physical states, and even their prediction.

In the 1940s the computers in Los Alamos were, in the true sense of the word, made of soldiers. Gathered in large groups, everyone had to solve a certain calculation step, but always the same step for the same man. It was here that computer development sought a revolution. The Monte Carlo Simulation, which originated at that time, was applied to the prediction of physical states of gas particles. From that time also the first applications of mechanical analogies on molecular systems were developed. The force fields were born and optimized and, in the course of time have achieved the unbelievable efficiency of modern times.

Mathematical approximation techniques have now made possible the quantum chemical calculation of systems even larger than the hydrogen atom, permitting "quantum dynamic" simulations of ligand binding at the active site of enzymes.

1.2
Do Today's Molecular Modeling Methods Illustrate only the Lukretian World?

This is in fact a question of quality of use. The methods could be used naively or intelligently, though the results are clearly distinguishable. However, naive uses should not be condemned, as it is vital for the quality of the use that a sufficiently critical position is taken when examining the results. In other words, the user realizes his or her naive use of the methods. Now, the researcher is conscious of the restrictions of the method and knows how to judge the results. Here, even with a very simple approach, this critical position results in further knowledge of the correlation between structure and properties.

Often however, such a critical attitude is not present—perhaps the result of modern commercial modeling systems. Those programs always provide a result, the evaluation of which is at liberty of the user. The programs tend stubbornly to calculate every absurd application and present a result—not only a number, but also a graph—and represent a further instrument of seduction for the uncritical use of algorithms. In contrast, the merits of molecular graphics is undisputed because of their essential contribution to the development of other analytical methods such as nuclear magnetic resonance spectroscopy and the X-ray analysis of proteins.

The tendency to perfect data presentation is the reverse situation. For example, visualization of isoelectrical potentials is one of the most valuable means of comparing molecular attributes. Very often a positive and a negative potential of a certain energy is used to describe structures. The presentation of potentials is based upon a charge calculation and may be used to find a suitable alignment of a training set of biologically active molecules. The latter can be realized on quite different quality levels. There are, for example, algorithms that perform well in calculations for simple carbohydrates, but are incapable of handling aromatic structures. Unfortunately these algorithms do not always signal their incapability if an aromatic system is to be calculated. A result is obtained, an isopotential surface is calculated, and a graph created. With that, an attempt is made to derive structure–activity relationships— the second trap comes next.

The training set that is selected, represents of course a drastic reduction of the parameter space. You may hope to receive a most possible representative distribution of the attributes by careful selection, but you are never sure. Thus, the correlations originate from the coincidential reciprocal completion of two errors, which relate back to the uncritical selection of methods and data sets.

1.3
What are Models Used for?

Models in science have different natures. They serve first of all to *simplify*, that means limiting of analysis to the phenomena that are believed to be the most important. Secondly, models serve as *didactical illustration* of very complicated circumstances, which are not easily accessible. Here, it must be taken into account in any explanation that the model does not show complete reality. A third model is that of *mechanical analogies*. These benefit from the fact that the laws of classical mechanics are completely defined, for example Hooke's law.

Model building of this kind plays a decisive role in the development of uniform theories. It is their special feature that it is not presupposed that the models reflect reality, but that first of all a structural similarity of two different fields is supposed. This is, for example, the presumption that the behavior of bonds in a molecule corresponds partly to springs, as described by Hooke's law. These mechanical analogy models have very successfully expanded theories, because the validity of a theory can in many cases be scrutinized experimentally, but the most important point is that predictions of new phenomena can be made.

These models are also often called *empirical*. Force fields belong to this class. The benefit of empiric models is that their parameters are optimized on reality. The "mechanization" does not provide explicit information from the non-mechanical contributions, but by empirical correction the non-mechanical contributions are convoluted in some way. That is why empirical models often are very close to reality.

Finally, the fourth type of model is *mathematical modeling*. These models serve for the simulation of processes, as for instance the kinetic simulation of a chemical reaction step in an enzyme. By suitable choice of parameters, kinetic simulations of real processes can be performed.

1.4
Molecular Modeling Uses All Four Types for Model Building

Didactical models are used for the combined representation of structure and molecular properties. In the case of *small molecules* the graphical representation of results from quantum chemical calculations or from the representation of the mobility of flexible ligands such as peptides. In the case of *proteins*, the structure itself is already a complex problem. Interactions of ligand and protein can also be studied with didactical models. It is already clear, that the different types of models are overlapping. Mechanical analogies, as well as reductions, aim at simplifying essential parts of the objects under study and are typical applications of molecular modeling.

1.5
The Final Step is *Design*

Design is perhaps the most essential element of all. Molecular modeling creates its own world, which is connected with reality by one of the four model types. Within this world—which exists in the computer—extrapolations can be made because, in contrast to the "real" world, a completely deterministical universe is created. Based on the analytical description of the system, the possibility is available to design inhibitors in advance of the synthesis and for them to be tested in a virtual computer experiment.

With that final design step, the circular course of a scientific study is completed. The study does not simply remain an analytical description of a system, which has been devised in "clockwork" fashion, but goes further by reassembling the system's parts. Molecular design creates a realization for our understanding—that a system could be more than simply the sum of its parts. This is especially effective for biological systems within which drug design is confronted by preference.

The design step itself actually is not as straightforward, even in the virtual world, as would be desirable. As *Gulliver* learns on his visit to the academy of Lagado, there is a machine, which at some time will have written every important scientific book of the world by a systematic combination of letters and words. *Jonathan Swift's* wonderful science fiction of the 18th century gives us at once the main problem: the time span of human beings is not large enough to test all possibilities. There has to be an intelligent algorithm to obtain the correct solutions. In the case of *Gullivers Travels,* Swift is somebody who introduces an additional criterion of quality. This is based on knowledge, experience, and is able to reject combinations of words and sentences: the human–machine network. Actually, Swift introduces such a criterion in the person of the professor who gives orders to his students, who serve the machines and decides after every experiment upon the result, e.g. lets the combination of words enter the book. Unfortunately, the experimenter himself is not defined qualitatively in Swift's novel; that is Swift's irony in *Gullivers Travels.* Hence, the result depends not only on an error-free function of the machine, but on the quality of its user! (Fig. 1.1)

The same problem is presented to us in the artificial world of modeling. Systematic exploration of properties is only possible for small numbers. Because of the combinatorics the system "explodes" after only a few steps. Flexibility studies on peptides give us a correct example. The change from four torsion angles to five or six increases the number of possible conformations from some thousands to several billions.

For the design of a ligand the situation becomes more complex. It demands a most intelligent restraint by suitable experiments, intuition or knowledge. Here also the quality of the human–machine network plays a decisive role. Fully automatic design systems seem to be like a Swift prediction machine in Gulliver's visit to the academy of Lagoda.

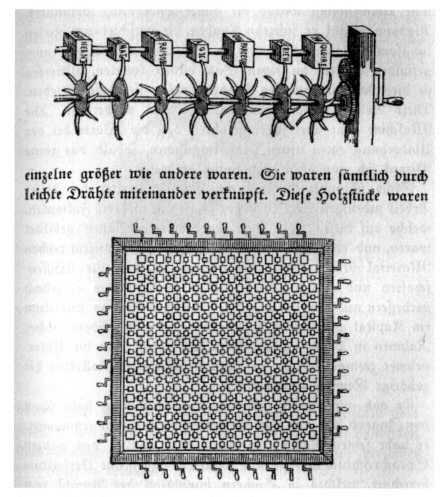

einzelne größer wie andere waren. Sie waren sämtlich durch leichte Drähte miteinander verknüpft. Diese Holzstücke waren

Fig. 1.1 J. J. Grandville's imaginary concept of the "book-writing machine" in Gulliver's visit to Lagoda.

1.6
The Scope of the Book

The scope of this book is to provide support for the beginner. The recognition of principal concepts and their limitations is important to us—more important even than a complete presentation of all available algorithms, programs and data banks. As with all areas associated with computer techniques the technical development in this area has been more than exponential. Almost every day, new algorithms are offered on the network, suitable for comparison of protein sequences or for searching of new data banks, etc. The user has no other possibility to judge their quality than to use the programs and to explore their limitations.

He or she must know, therefore, that energy-minimizing in vacuo does not make sense in any case for the analysis of the interaction geometry of a ligand. He or she also has to know that a protein can not be simply folded up from a linear polypeptide chain. It must be realized that there is alternative or multiple binding mode: inhibitors binding to an enzyme show alternative binding geometries in the active site, even within a set of analogs. Very small changes in the molecular structure could provoke another orientation of the ligand in the active site. It is not necessarily true that a structure-orientated superposition would be better than an intuitive one or even one, which is oriented by steric or electrostatic surface properties.

Today's modeling in essence goes far beyond the example of Lukrez. Modeling is no longer on the level of analytic description of properties or correlations. It is much more than the creation of "colored pictures"—it also introduces us to systematic thinking. It even *demands* systematic thinking in order to avoid too many simple applications and to keep in mind the limitations of the methods.

Here we also want to provide support. By describing our own experiences with molecular design in two examples, one for "small molecules" (ligands) and another for "big molecules" (proteins), we aim to encourage the beginner to a critical engagement, hopefully without demotivation.

2
Small Molecules

2.1
Generation of 3D Coordinates

When starting a molecular modeling study the first thing to do is to generate a model of the molecule in the computer by defining the relative positions of the atoms in space by a set of cartesian coordinates. A reasonable and reliable starting geometry essentially determines the quality of the following investigations. It can be obtained from several sources. The four basic methods for generating 3D molecular structures are

1. use of X-ray crystallographic databases,
2. compilation from fragment libraries with standard geometries, and
3. simple drawing of 2D-structures using an approach called 'sketch',
4. 2D to 3D conversion using automated approaches.

2.1.1
Crystal Data

First we will focus on the use of X-ray data for molecular building. The most important database for crystallographic information studying small molecules is the Cambridge Crystallographic Database [1]. This database contains experimentally derived atomic coordinates for organic and inorganic compounds up to a size of about 500 atoms and is continuously updated. The Cambridge Crystallographic Data Centre leases the database as well as software for searching the database and for analyzing the results. The output of the database search is a simple, readable file containing the 3D-structural information about the molecule of interest. This data file can be read by most of the commercial molecular modeling packages [e.g. 2, 3].

The atomic coordinates listed in the database are converted automatically to cartesian coordinates when reading the file into the modeling program. Subsequently the structure can be displayed by molecular graphics and studied in its 3D shape.

In general, small molecule X-ray structures are very well resolved but there is no guarantee for the accuracy of the data. The localization of hydrogen atoms always is a problem because they are difficult to observe by X-ray crystallography. The principle of the X-ray method is the scattering of the X-rays by the electron cloud around an atom. Because hydrogen atoms have only one electron, their influence on X-ray scattering is low and they are normally disregarded in structure determination. But of course hydrogen positions can be appointed on the basis of collected knowledge on standard bond lengths and bond angles. According to this procedure all bond lengths involving hydrogen atoms are usually not very specific. Before using the information from the X-ray database it is therefore advisable to check the atomic coordinates, bond lengths and bond angles for internal consistency. The following points especially should be clarified before starting any work with a X-ray structure:

1. are the atom types correct
2. are the bond lengths and bond angles reasonable
3. are the bond orders correct

and in case of chiral molecules,

4. do the data correspond to the correct enantiomer?

After taking care of these details the molecule can be saved in a molecular data file. The organization, extension name, format, and the information contained in the file are program-dependent.

It should be kept in mind that the crystalline state geometry of a molecule is subject to the influence of crystal packing forces. Therefore bond lengths and bond angles can differ from theoretical standard values. Furthermore, the solid state structure corresponds to only one of perhaps many low-energy conformations accessible to a flexible molecule and is always affected by the neighbor molecules in the crystal unity cell and sometimes also influenced by solvent molecules in the crystal. Other energetically allowed conformations must be explored by a conformational analysis eventually to reveal conformations of biological relevance. Also, knowledge of the most stable conformation called the *global energy minimum structure* is important to allow the evaluation of probabilities for conformers with higher energy content. Procedures for this purpose are described in section 2.2.

2.1.2
Fragment Libraries

The second very common building method is the construction of molecules from pre-existing fragment libraries. This is the method of choice when there is no access to crystallographic databases or if X-ray data for the desired structures are not available. Almost all commercial molecular modeling programs nowadays offer the possibility to construct molecules using fragment libraries.

Fragment libraries can be utilized like an electronic 3D structure tool kit, which is easy to handle. Because of the preoptimized standard geometries of all entries in the fragment pool resulting 3D structures already have an acceptable geometry. In most cases only torsion angles have to be cleared to avoid atom overlapping or close van der Waals contacts. Problems may arise with fused ring systems because of the different ways in which saturated rings can be joined to each other. To solve this pro-

blem it is recommended wherever possible to refer to X-ray data or to experimental data of comparable ring systems in order to select the correct ring connection.

Each atom in any arbitrary structure carries characteristic features which are defined by the so-called atom type. Properties distinguishing between different atoms in molecular modeling terms are for example hybridization, volume, etc. The corresponding parameters define the particular atom type. All atomic parameters taken collectively represent the atomistic part of a force field. On pre-existing fragments selected from libraries the atom types of course are already defined and in general are correct. In many cases, however, the decision as to which atom type will be appropriate is less easy to take. We will discuss this problem on the example of N-acetylpiperidine.

If N-acetylpiperidine is generated from the fragment library using a piperidine ring and an acetyl residue the nitrogen atom in the piperidine is defined as sp3 nitrogen atom type with tetrahedral geometry. But if this nitrogen is connected with the acetyl residue it also can be considered as an amide nitrogen atom demanding planar trigonal sp^2 geometry. In such a case the correct decision can only be made by either comparing the geometry obtained from the building routine with X-ray data or performing a quantum mechanical calculation for the structural element of interest in order to get a reliable geometry. Fig. 2.1 shows the results of a semiempirical and an ab initio calculation in comparison with force field geometries and the crystal structure of N-acetyl-piperidine-2-carboxylic-acid [4].

While the sp^3 nitrogen atom of the force field structure bears a tetrahedral geometry the crystal structure and the quantum chemically calculated geometries indicate an almost planar nitrogen atom. To avoid errors in subsequent calculations the nitrogen atom has to be assigned an atom type with planar geometry.

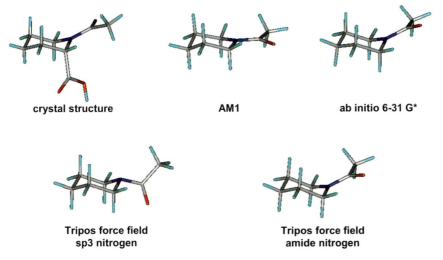

| crystal structure | AM1 | ab initio 6-31 G* |

| Tripos force field sp3 nitrogen | Tripos force field amide nitrogen |

Fig. 2.1 The geometry of the amide group in N-acetyl piperidine depends crucially on the method used as well as the atom types employed for optimization. For comparison the crystal structure of piperidine-2-carboxylic-acid is shown in the upper left. The color code: carbon = white, oxygen = red, nitrogen = blue, hydrogen = cyan, sulfur = yellow, halogens = green, is used throughout this book.

Another problem occuring when building substituted saturated ring systems is the correct conformation of the cycle, because it may be influenced by the substituents. Cyclohexane is one of the most detailed studied cyclic molecules in organic chemistry. The different possible conformations and the energy barriers separating them have been the subject of many investigations [5, 6]. There is no doubt that the chair form is the most stable conformation of this molecule. For monosubstituted cyclohexane this still holds true. The preferred position of any substituent is found to be the equatorial one. The energy difference determined between the equatorial and axial position is not very distinct for small substituents but is increasing for larger groups [7]. Therefore it is necessary and advisable always to check the results of structure building from fragment libraries in comparable situations with experimental data.

2.1.3
Sketch Approach

The simplest method of structure generation is the so-called sketch approach. When using this routine the mouse pointer functions as a simple pencil to draw a 2D formula of the molecule on the computer screen. Sometimes a very limited number of small molecular standard fragments is already available from a library and can be used as starting points. When finishing the drawing process the 2D picture on the screen is converted into 3D information. Because of this procedure the setting of correct atom types should be watched especially carefully. Since the sketch approach is a very simple method the resulting geometries in general are not very satisfying. Therefore a rough geometry optimization is performed automatically at the end of each sketch operation in order to relax the molecular geometry.

2.1.4
Conversion of 2D Structural Data into 3D Form

An alternative way for generating three-dimensional molecular structures is to start from 2D or 2.5D representations of molecules and to convert this information into a 3D form. While in the already mentioned sketch approach a single formula drawing is converted into three-dimensional information, programs like CONCORD [8, 9] and CORINA [10] offer the possibility to automatically generate 3D structures. Both programs use a rule- and data-based system that automatically generates 3D atomic coordinates from the constitution of a molecule as expressed by a connection table or linear code, and that is powerful and reliable enough to convert large databases of several hundreds of thousands or even millions of compounds.

CONCORD was specifically developed for the 2D to 3D conversion of large database entries containing connection tables of potentially bioactive molecules. For structure generation CONCORD uses a very detailed table of bond lengths. In addition to data like atomic number, hybridization state and bond type the program takes into account the 'environment' of the atoms included in the bond before assigning the bond lengths. This precise selection of bond lengths is especially important for the construction of ring systems. Deviations from correct bond lengths may have a dramatic effect on the resulting ring conformation.

At the beginning of the 2D conversion the program identifies the so called 'smallest set of small rings'. Subsequently a logical analysis is performed for each particular ring system. Based upon ring adjacency and ring constraints these logical rules decide how the rings will be constructed. In addition a rough conformation of each ring system is determined taking into consideration planarity and stereochemical constraints.

If fusion atoms of multicyclic systems are not specified, CONCORD creates the isomer with the lowest energy content. After constructing and connecting the ring systems the program modifies the gross conformations in order to remove the internal strain by distributing the strain symmetrically over all atoms in the ring. This procedure leads to cyclic structures with sufficiently relaxed geometries.

The next step in structure generation is to add the acyclic substructures. Bond lengths and bond angles again are taken from predefined tables. To avoid close van der Waals contacts in the calculated structure the torsion angles are modified in order to obtain energetically acceptable conformations. Besides computational speed the main advantage of CONCORD is that the entire topology of the growing molecule is considered at each step. As a result of this CONCORD yields 3D-structures of good quality at low demands of computer time. This is an important criterion when large databases of 2D information are to be converted into 3D space.

CORINA works very similar to CONCORD. The starting point in creating ring systems is analogous to CONCORD. But in the following step CORINA uses a different approach to connect the ring systems. The rings are fused and the energies of possible ring conformations are calculated using a crude force field. If the current choice of a particular ring connection is found to be energetically unfavorable, a new attempt is made using other energetically possible conformations of the rings. The generation of ring structures is followed by a geometry optimization step.

Similar to CONCORD, the acyclic substructures are constructed after the ring system is completed. The chains added to the rings are usually in fully extended conformations. This, of course, leads to geometries needing further refinement. The torsion angles are rotated until the first conformation is reached that avoids close contacts. As a result of this rough conformational search the program indeed yields acceptable structures.

It is important to note that the resulting conformations will only as a matter of chance correspond to either a conformation in the crystal environment or to a low energy conformation. The finally obtained structure, therefore, has to be subjected to a conformational analysis in order to detect all possible low energy conformations.

Both of the reviewed programs are effective alternatives in structure generation. They are fast (less than 1 second for small and medium-sized organic molecules on a common workstation), robust and provide good conversion rates (99.5%) as tested on the conversion of 250,000 2D structures of the National Cancer Institute Open Database. This database is freely available and contains a huge number of small organic compounds and drug molecules tested for cancer activity at the National Cancer Institute.

References

[1] Cambridge Structural Database, Dr. Olga Kennard, F.R.S., Cambridge Crystallographic Data Centre, 12 Union Road, Cambridge CB2 1EZ, U.K.

[2] SYBYL, Tripos Associates, St. Louis, Missouri, USA.

[3] INSIGHT/DISCOVER, Accelrys Inc., San Diego, California, USA.

[4] Rae, I. D., Raston, C. L., and White, A. H. *Aust. J. Chem.* **33**, 215 (1980).

[5] Bucourt, R. The torsion angle concept in conformational analysis. In: *Topics in Stereochemistry*, Vol. 8. Eliel, E. L., and Allinger, N. L. (Eds.). Wiley: New York; 159–224 (1974).

[6] Shopee, C. W. *J. Chem. Soc.* **II**, 1138–1151 (1946).

[7] Hirsch, J. A. Tables of conformational energies. In: *Topics in Stereochemistry*, Vol. 1. Eliel, E. L., and Allinger, N. L. (Eds.). Wiley: New York; 199–222 (1967).

[8] Pearlman, R. S. *CDA News* **2**, 1–7 (1987).

[9] Pearlman, R. S. 3D Molecular Structures: Generation and Use in 3D Searching. In: *3D QSAR in Drug Design*. Kubinyi, H. (Ed.). Escom Science Publishers: Leiden; 41–79 (1993).

[10] Gasteiger, J., Rudolph, C. and Sadowski, *J. Tetrahedron Comput. Methodol.* **3**, 537–547 (1990).

2.2
Computational Tools for Geometry Optimization

2.2.1
Force Fields

Molecular structures generated using the procedures described in the previous section should always be geometry optimized to find the individual energy minimum state. This is normally done by applying a molecular mechanics method. The expression "molecular mechanics" is used to define a widely accepted computational method employed to calculate molecular geometries and energies.

Unlike quantum mechanical approaches the electrons and nuclei of the atoms are not explicitly included in the calculations. Molecular mechanics considers the atomic composition of a molecule to be a collection of masses interacting with each other via harmonic forces. As a result of this simplification molecular mechanics is a relatively fast computational method practicable for small molecules as well as for larger molecules and even oligomolecular systems.

In the framework of the molecular mechanics method the atoms in molecules are treated as rubber balls of different sizes (atom types) joined together by springs of varying length (bonds). For calculating the potential energy of the atomic ensemble use is made of Hooke's law. In the course of a calculation the total energy is minimized with respect to atomic coordinates where:

$$E_{tot} = E_{str} + E_{bend} + E_{tors} + E_{vdw} + E_{elec} + ... \tag{1}$$

where E_{tot} is the total energy of the molecule, E_{str} is the bond-stretching energy term, E_{bend} is the angle-bending energy term, E_{tors} is the torsional energy term, E_{vdw} is the van der Waals energy term, and E_{elec} is the electrostatic energy term.

Molecular mechanics enables the calculation of the total steric energy of a molecule in terms of deviations from reference "unstrained" bond lengths, angles and torsions plus non-bonded interactions. A collection of these unstrained values, together with what may be termed force-constants (but in reality are empirically derived fit parameters), is known as the *force field*. The first term in Eq. (1) describes the energy change as a bond stretches and contracts from its ideal unstrained length. It is assumed that the interatomic forces are harmonic so the bond-stretching energy term can be described by a simple quadratic function given in Eq. (2):

$$E_{str} = \frac{1}{2} k_b (b-b_0)^2 \tag{2}$$

where k_b is the bond-stretching force constant, b_0 is the unstrained bond length, and b is the actual bond length.

In more refined force fields a cubic term [1], a quartic function [2–4], or a Morse function [5] has been included.

Also for angle bending mostly a simple harmonic, spring-like representation is employed. The expression describing the angle-bending term is shown in Eq. (3):

$$E_{bend} = \frac{1}{2} k_\theta (\theta - \theta_0)^2 \tag{3}$$

where k_θ is the angle-bending force constant, θ_0 is the equilibrum value for θ, and θ is the actual value for θ.

A common expression for the dihedral potential energy term is a cosine series, as Eq. (4):

$$E_{tors} = \frac{1}{2} k_\varphi (1 + \cos(n\varphi - \varphi_0)) \tag{4}$$

where k_φ is the torsional barrier, φ is the actual torsional angle, n is the periodicity (number of energy minima within one full cycle), and φ_0 is the reference torsional angle (the value usually is 0° for a cosine function with an energy maximum at 0° or 180° for a sine function with an energy minimum at 0°).

The van der Waals interactions between not directly connected atoms are usually represented by a Lennard-Jones potential [6] (Eq. 5).

$$E_{vdw} = \sum \frac{A_{ij}}{r_{ij}^{12}} - \frac{B_{ij}}{r_{ij}^6} \tag{5}$$

where A_{ij} is the repulsive term coefficient, B_{ij} is the attractive term coefficient, and r_{ij} is the distance between the atoms i and j.

This is one form of the Lennard-Jones potential but there exist several modifications of this term used in the different force fields. An additional function is used to describe the electrostatic forces. In general it is made use of the Coulomb interaction term (Eq. 6).

$$E_{elec} = \frac{1}{\varepsilon} \frac{Q_1 Q_2}{r} \tag{6}$$

where ε is the dielectric constant, Q_1, Q_2 are atomic charges of interacting atoms, and r is the interatomic distance.

Charges may be calculated using the methods described in section 2.4.1.1 or are implemented in some of the force fields [2–4] as empirically derived parameter sets.

Some force fields also include cross terms, out of plane terms, hydrogen bonding terms etc. and use more differentiated potential energy functions to describe the system. As force fields are varying in their functional form not all can be discussed here in detail but they have been subject of excellent reviews [7, 8].

The basic idea of molecular mechanics is that the bonds have "natural" lengths and angles. The equilibrium values of these bond lengths and bond angles and the corresponding force constants used in the potential energy function are defined in the force field and will be denoted as *force field parameters*. Each deviation from these standard values will result in increasing total energy of the molecule. So, the total energy is a measure of intramolecular strain relative to a hypothetical molecule with ideal geometry. By itself the total energy has no physical meaning.

The objective of a good and generally employable force field is to describe as many as possible different classes of molecules with reasonable accuracy. The reliability of the molecular mechanics calculation is dependent on the potential energy functions and the quality of the parameters incorporated in these functions. So, it is easy to understand that a calculation of high quality can not be performed if parameters for important geometrical elements are missing. To avoid this situation it is necessary to choose a suitable force field for a particular investigation.

Several force fields have been developed to examine a wide range of organic compounds and small molecules [1–4, 9] while other programs contain force fields primarily for proteins and other biomolecules [10–12]. If parameters for particular atom types are missing it is unavoidable to add the missing data to the force field [13–15].

2.2.2
Geometry Optimization

As already mentioned almost certainly the generated 3D model of a given molecule does not have ideal geometry; therefore, a geometry optimization must be performed subsequently. In the course of the minimization procedure the molecular structure will be relaxed. As can be deduced from the example presented in Fig. 2.2 and Table 2.1 the internal strain in structures obtained from crystal data is mainly influenced by small deviations from the "ideal" bond lengths. Therefore above all the corresponding energy terms (bond-stretching term, angle-bending term) are altered in course of a force field optimization. Despite the remarkable change in energy content torsional angles are effected only to a lesser extent. This is a clear indication to the well-known observation that in crystals almost exclusively low-energy conformations are found. It also should be realized that crystal structures are by no means "bad" geometries. As can be easily deduced from Fig. 2.2 the distortion of the crystal structure when compared with the relaxed geometry of the force field structure in terms of geometry differences is only very subtle. This fact can be inter-

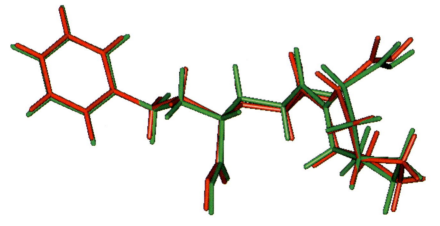

Fig. 2.2 Superposition of the crystal structure (red) and force-field-optimized geometry (green) of the angiotensin-converting enzyme inhibitor ramiprilate.

Tab. 2.1 Force field energy terms for the ramiprilate molecule before and after geometry optimization

Structure	Energy (kcal mol^{-1})
Crystal	
Bond-stretching energy	179.514
Angle-bending energy	15.693
Torsional energy	17.230
Out-of-Plane-bending energy	0.043
1–4 van der Waals energy	18.538
van der Waals energy	−3.839
Total energy	227.178
Optimized	
Bond-stretching energy	0.982
Angle-bending energy	10.372
Torsional energy	14.335
Out-of-Plane-bending energy	0.011
1–4 van der Waals energy	4.791
van der Waals energy	−7.822
Total energy	22.669

preted also in the sense that large variations in geometry are not to be expected when different well-parameterized force fields are applied. In the case considered here the individual but real crystal packing of ramiprilate is compared to the well-known general Tripos force field.

Before starting a geometry optimization, bad van der Waals contacts should be removed because the minimum energy geometry at the end of the optimization will depend on the starting geometry [7].

Several advantages like speed, sufficient accuracy and the broad applicability on small molecules as well as on large systems have established the force field geometry optimization as the most important standard method. Because of the complexity and the demanding computational costs quantum mechanical methods should be reserved for special problems which will be discussed later.

We will now focus on some common energy minimization procedures used by molecular mechanics. It is important to note that the minimization algorithms only find local minima on the potential energy surface but not implicitly the global energy minimum.

2.2.3
Energy-Minimizing Procedures

The energy minimization methods can be divided into two classes: the first-derivative techniques like steepest descent, conjugate gradient and Powell; and the second-derivative methods like the Newton–Raphson and related algorithms.

2.2.3.1 Steepest Descent Minimizer

The steepest descent minimizer uses the numerically calculated first derivative of the energy function to approach the energy minimum. The energy is calculated for the initial geometry and then again when one of the atoms has been moved in a small increment in one of the directions of the coordinate system. This process will be repeated for all atoms which finally are moved to new positions downhill on the energy surface [7]. The procedure will stop if the predetermined minimum condition is fulfilled. The optimization process is slow near the minimum, so the steepest descent method is often used for structures far from the minimum. It is the method most likely to generate low-energy structures of poorly refined crystallographic data or to relax graphically built molecules. In most cases the steepest descent minimization is used as a first rough and introductory run followed by a subsequent minimization employing a more advanced algorithm like conjugate gradients.

2.2.3.2 Conjugate Gradient Method

The conjugate gradient method accumulates the information about the function from one iteration to the next. With this proceeding the reverse of the progress made in an earlier iteration can be avoided. For each minimization step the gradient is calculated and used as additional information for computing the new direction vector of the minimization procedure. Thus, each successive step continually refines the direction towards the minimum. The computational effort and the storage requirements are greater than for steepest descent but conjugate gradients is the method of choice for larger systems. The greater total computational expense and the longer time per iteration is more than compensated by the more efficient convergence to the minimum achieved by conjugate gradients.

The Powell method is very similar to conjugate gradients. It is faster in finding convergence and is suitable for a variety of problems, but one should be careful when using the Powell algorithm because torsion angles may sometimes be modified to a dramatic extent. So, the Powell method is not practicable for energy minimization after a conformational analysis because the located low-energy conformations will be altered in an undesired manner. It is advisable to perform a conjugate gradient minimization in this situation.

2.2.3.3 Newton–Raphson Minimizer

The Newton–Raphson minimizer as a second-derivative method uses, in addition to the gradient, the curvature of the function to identify the search direction. The second derivative is also applied to predict where the function passes through a minimum. The efficiency of the Newton–Raphson method increases as convergence is approached. The computational effort and the storage requirements for calculating larger systems are disadvantages of this method. For structures with high strain the minimization process can become instable, so the application of this algorithm is mostly limited to problems where rapid convergence from a preoptimized geometry to an extremely precise minimum is required. For some more detailed information about the optimization methods see [16, 17].

It can be summarized that the choice of the minimization method depends on two factors—the size of the system and the current state of the optimization. For structures far from minimum, as a general rule, the steepest descent method is often the best minimizer to use for the first 10–100 iterations. The minimization can be completed to convergence with conjugate gradients or a Newton–Raphson minimizer. To handle systems that are too large for storing and calculating a second-derivative matrix the conjugate gradient minimizer is the only practicable method. The minimization procedure will continue until convergence has been achieved.

There are several ways in molecular minimization to define convergence criteria. In non-gradient minimizers like steepest descent only the increments in the energy and/or the coordinates can be taken to judge the quality of the actual geometry of the molecular system. In all gradient minimizers, however, atomic gradients are used for this purpose. The best procedure in this respect is to calculate the root mean square gradients of the forces on each atom of a molecule. It is advisable also always to check the maximum derivative in order to detect unfavorable regions in the geometry. There is no doubt about the quality of a minimum geometry if all derivatives are less than a given value. The specific value chosen for example for the maximum derivative depends on the objective of the minimization. If a simple relaxation of a strained molecule is desired, a rough convergence criterion like a maximum derivative of 0.1 kcal mol^{-1} Å^{-1} is sufficient while for other cases convergence to a maximum derivative less than 0.001 kcal mol^{-1} Å^{-1} is required to find a final minimum.

The choice of the convergence criteria should be a balance between attaining reasonable accuracy in determining the minimum structure and avoiding unnecessary computations when no further progress can be realized [17].

2.2.4
Use of Charges, Solvation Effects

Molecular mechanics calculations are usually carried out under vacuum conditions ($\varepsilon = 1$). For unpolar hydrocarbons the effect of the explicit inclusion of solvent as compared with gas phase calculations is negligible. The investigation of molecules containing charges and dipoles however requires the consideration of solvent effects [7]; otherwise conformations mainly influenced by strong electrostatic interactions would be overestimated. The force field will try to maximize the attractive electrostatic interaction, resulting in energetically strongly preferred but unrealistic low-energy conformations of the molecule. This can be prevented by employing the corresponding solvent dielectric constant [18]. For example, in water ε amounts to 80. In contrast to macromolecules, the electrostatic field of small molecules is considered to be homogeneous; therefore the use of an uniform dielectric constant in principle is allowed. Experimentally determined dielectric constants for a large number of solvents may be found in the literature and can be applied for a correct treatment of the Coulombic term of solvated molecules.

A very simple but effective way to treat the problem of charges and solvation in the course of a molecular mechanics optimization is to perform the calculation without taking charges into account. This very often yields acceptable results and is especially recommended if the results of a conformational analysis are to be minimized because usage of charges may markedly alter the conformation by electrostatic interactions. Consideration of charges always is necessary if hydrogen bonding phenomena are to be described.

The strength of the electrostatic interaction decreases with r^{-1}. Therefore, in some force fields the dielectric constant can be chosen to be distance-dependent in order to simulate the effect of displacement of solvent molecules in course of the approach of a ligand molecule to a macromolecular surface. This is of particular value if a conformational analysis is part of a pharmacophore search.

Whenever possible experimental data should be used for testing results from theoretical calculations. Above all, NMR data have become a valuable tool in this respect. Since most of the available NMR data have been obtained in chloroform or similar organic solvents, the explicit inclusion of the corresponding dielectric constant in the Coulombic term of a force field leads to an improved agreement with experimental results.

Consideration of the dielectric constant is one possibility to simulate solvent effects. An alternative way is to create a solvent box around the molecule containing discrete solvent molecules. The additional computational effort and the limitations in regard to the limited number of solvents that can be used in most of the available force fields are severe disadvantages of this method.

2.2.5
Quantum Mechanical Methods

Quantum mechanical methods also must be discussed, at least in brief, because they are very valuable additional tools in computational chemistry. In general, properties like molecular geometry and relative conformational energies can be calculated with high accuracy for a broad variety of structures by a well-parametrized general force field. However, if force field parameters for a certain structure are not available quantum chemical methods can be used for geometry optimization. In addition, the calculation of transition states or reaction paths as well as the determination of geometries influenced by polarization or unusual electron distribution in a molecule is the domain of quantum mechanical calculations. Their disadvantages relative to other methods are the computational costs and the limitation to rather small molecules. So, the use of quantum mechanical methods should be reserved for the treatment of special problems. The objective in this context is not to discuss the quantum mechanical methods from a theoretical perspective but to give some practical hints for the application of semiempirical or ab initio programs. The reader's interest may be drawn to many books and reviews on this subject to gain more insight into the theoretical aspects of these methods [19–22].

2.2.5.1 **Ab initio Methods**

Unlike molecular mechanics and semiempirical molecular orbital methods ab initio quantum chemistry is capable of reproducing experimental data without employing empirical parameters. Therefore, the application of ab initio calculations is especially favored in situations in which little or no experimental information are available.

The quality of an ab initio calculation depends on the basis set used for the calculation [23, 24] and the computational method employed. A wrong choice of the basis set can render the results of extremely time-consuming calculations meaningless. The decision which basis set should be used is related to the objective of the calculation and the molecules to be studied. It should be kept in mind that even a large basis set is not always a guarantee for agreement with experimental data [25].

Only the most commonly applied basis sets will be discussed here. The STO-3G (Slater type orbitals approximated by three Gaussian functions each) basis set has been frequently used in the past and is the smallest basis set that can be chosen. This minimal basis set consists of the smallest number of atomic orbitals necessary to accommodate all electrons of the atoms in their ground state, assuming spherical symmetry of the atoms.

In more recent ab initio calculations the split-valence basis sets have become quite popular. In these the valence orbital shells are represented by an inner and outer basis function. In this way more flexibility in describing the residence of the electrons has been attained [26]. The split-valence basis sets represent a progress over the STO-3G basis set, and the 3–21G, 4–31G, and 6–31G basis sets are widely used in ab initio calculations. They differ only in the number of primitive Gaussians used in expanding the inner shell and first contracted valence function [25]. 4–31G for example means that the core orbitals consist of four and the inner and outer valence orbitals of three and one Gaussian functions, respectively.

The next level of improvement is the introduction of polarization basis sets. To all non-hydrogen atoms d orbitals are added to allow p orbitals to shift away from the position of the nucleus leading to a deformation (polarization) of the resulting orbitals. This adjustment is particulary important for compounds containing small rings [26]. The polarization basis sets are marked by a star, e.g. 6–31G*. This basis set uses six primitive Gaussians for the core orbitals, a three/one split for the s and p valence orbitals, and a single set of six d functions (indicated by the asterisk).

For a more detailed description of the basis sets the reader is directed to books and reviews on this subject [22, 25].

Unfortunately there is no general rule for choosing an adequate basis set. The level of calculation depends on the desired accuracy and the molecular properties of interest. A geometry optimization of a simple molecule with moderate size reasonably can be performed using a 3–21G basis set. For other problems, however, this degree of sophistication may not be sufficient. If the geometry of the molecule is influenced by polarization effects, electron delocalization or hyperconjugative effects a 6–31G* or higher basis set is necessary to include the d orbitals as already mentioned (Fig. 2.3).

In spite of the rapid development in computer technology, high level ab initio calculations still can not always be performed. A common way to overcome the pro-

STO 3G	**3-21 G**	**6-31 G***

Fig. 2.3 This shows the final geometries of 2,6-diazaspiro[3.3]hept-2-yl-formamide after geometry optimization using different basis sets. The example clearly indicates the dependence of the resulting geometry on the applied basis set. The minimal basis set STO-3G and the 3–21G basis set yield very different geometries. The inclusion of d orbitals (6–31G*) leads to a structure reflecting the polarization effects and the ring tension more precisely. The resulting geometry of the amide nitrogen atom lies between tetrahedral and trigonal planar hybridization states.

blem of excessive computational requirements is the use of a 3–21G basis set to optimize the geometry parameters and then to compute the wavefunction on the 6–31G* level. This procedure is often termed 6–31G*//3–21G calculation.

The use of higher basis sets does not automatically improve the accuracy of the calculated molecular properties of interest. In order to find a suitable level of calculation it is necessary to calibrate the method against experiment or testing the basis sets empirically to yield acceptable results.

2.2.5.2 Semiempirical Molecular Orbital Methods

The deep gap between molecular mechanics and the ab initio calculations is occupied by the semiempirical molecular orbital methods. They are basically quantum mechanical in nature but the main difference to ab initio methods is the introduction of empirical parameters in order to reduce the high costs of computer time necessary for explicit evaluation of all integrals. One-center repulsion integrals and resonance integrals are substituted by parameters fitted as closely as possible to experimenal data.

Another basic idea of the semiempirical approach is the consideration of the fact that most of the interesting molecular properties are mainly influenced only by the valence electrons of the corresponding atoms. Therefore only the valence electrons are taken into account, leading to a further reduction in computer time.

All the semiempirical methods apply the same theoretical assumptions, they only differ in the approximations beeing made [27]. Semiempirical methods like AM1 [28] and PM3 [29–31] provide a quite effective compromise between the accuracy of the results and the expense of computer time required. A calculation performed with AM1 or PM3 is able to reflect the experiment as effectively as an ab initio calculation using a small basis set. The advantage of semiempirical methods over ab initio calculations is not only that they are several orders of magnitude faster, but also that calculations for systems up to 200 atoms are possible with the semiempirical methods only. However, it is recommended to check one's results carefully. Like the choice of a wrong basis set in ab initio calculations, the lack of correct parameters in semiempirical studies can also lead to meaningless results. The quality of semiempirical methods for a wide range of molecules and the calculation of different properties has been subject of several reviews [28–31]. It should be noted that in general semiempirical methods may give erroneous results for the third-row elements.

2.2.5.3 Combined Quantum Mechanical (QM) and Molecular Mechanical (MM) Methods

The theoretical limitations of molecular mechanics as well as the computational intensity of quantum mechanical calculations lead to the development of a hybrid model combining the individual advantages of these two approaches, which was first introduced by Warshel and Levitt in 1976 [32].

Despite the enormous progress in computer technology and theoretical methodology that has increased significantly the number of atoms which can be treated quantum mechanically in recent years, it is still not possible to apply quantum mechanics on large biological molecules such as proteins, DNA and lipid membranes consisting of thousands of atoms. Unfortunately, the usage of quantum mechanics is indispensable to describe chemical reactions involving the breaking and formation of covalent bonds or to calculate realistic free energies for the binding of drugs to their targets. On the other hand, it is often not necessary to treat the whole macromolecule and solvent with quantum mechanics because the processes that need to be described in an electronically accurate fashion occur in strictly localized regions (e.g. the active site in an enzyme). This was the basis for the development of hybrid models where only the relevant part of a macromolecule and its potential ligands is treated quantum mechanically while the rest of the macromolecule and solvent is described with molecular mechanics. A schematic arrangement of such a QM/MM system is shown in Fig. 2.4. Exploring Fig. 2.4, one can easily understand why QM/MM methods are also called 'embedding' methods. The QM part is normally buried (embedded) in the MM region of a QM/MM system.

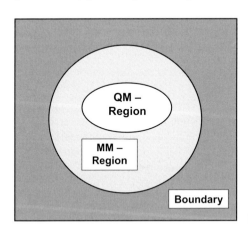

Fig. 2.4

The modifications that have to be performed in order to allow combination of a force field with an ab initio or semiempirical method are basically very similar in the different proposed theories. The different theories have to be combined so that QM- and MM-atoms can interact, i.e. QM-atoms can "see" MM-atoms (and the vice versa). This is made possible by incorporation of electrostatic interactions between QM- and MM-atoms in the QM energy calculation. Van der Waals interactions of QM- and MM-atoms are considered classically in an independent energy term. The total energy of a QM/MM approach can in principle be summarized as in equation (7).

$$E_{tot} = E_{QM,elec} + E_{QM,vdW} + E_{MM} \tag{7}$$

E_{tot}	=	total energy
$E_{QM,elec}$	=	electrostatic energy of QM-atoms
$E_{QM,vdW}$	=	van der Waals energy of QM-atoms
E_{MM}	=	energy of MM-atoms

In general any combination of a force field and a quantum mechanical program is possible. Some well-known and often-used software programs already offer a QM/MM module (CHARMM, QSite, QuanteMM [33–35]).

In recent years QM/MM potentials have been applied increasingly for the investigation of enzyme-catalyzed reactions. The description of enzyme reactions at the atomic level provides new insights and helps to understand enzymatic mechanisms. The first enzyme which was investigated by QM/MM calculations was lysozyme [32]. Other well-described enzymes are triosephosphate isomerase [36] and citrate synthase [37].

References

[1] Allinger, N. L. *J. Am. Chem. Soc.* **99**, 8127–8134 (1977).

[2] Allinger, N. L., Yuh, Y. H., and Lii, J.-H. *J. Am. Chem. Soc.* **111**, 8551–8566 (1989).

[3] Lii, J.-H., and Allinger, N. L. *J. Am. Chem. Soc.* **111**, 8566–8576 (1989).

[4] Lii, J.-H., and Allinger, N. L. *J. Am. Chem. Soc.* **111**, 8576–8582 (1989).

[5] Morse, P. M. *Phys. Rev.* **34**, 57 (1929).

[6] Lennard-Jones, J. E. *Proc. Roy. Soc.* **106A**, 463 (1924).

[7] Burkert, U., and Allinger, N. L. *Molecular Mechanics*. ACS Monograph 177. American Chemical Society: Washington D. C. 1982.

[8] Dinur, U., and Hagler, A. T. New Approaches to Empirical Force Fields. In: *Reviews in Computational Chemistry*, Vol. 2. Lipkowitz, K. B., and Boyd, D. B. (Eds.). VCH: New York; 99–164 (1991).

[9] Clark, M., Cramer III, R.D., and Van Opdenbosch, N. *J. Comput. Chem.* **10**, 982–1012 (1989).

[10] Dauber-Osguthorpe, P., Roberts, V.A., Osguthorpe, D.J., Wolff, J., Genest, M., and Hagler, A.T. *Proteins Struct. Func. Gen.* **4**, 31–47 (1988).

[11] Brooks, B. R., Bruccoleri, R. E., Olafson, B. D., States, D. J., Swaminathan, S., and Karplus, M. *J. Comput. Chem.* **4**, 187–217 (1983).

[12] van Gunsteren, W.F., and Berendsen, H.J.C. Molecular dynamics simulations: techniques and applications to proteins. In: *Molecular Dynamics and Protein Structure*. Hermans, J. (Ed.). Polycrystal Books Service: Western Springs, Illinois; 5–14 (1985).

[13] Hopfinger, A. J., and Pearlstein, R. A. *J. Comput. Chem.* **5**, 486–499 (1984).

[14] Maple, J.R., Dinur, U., and Hagler, A. T. *Proc. Natl Acad. Sci., U.S.A.* **85**, 5350–5354 (1988).

[15] Bowen, J. P., and Allinger, N. L. Molecular Mechanics: The Art and Science of Parameterization. In: *Reviews in Computational Chemistry*, Vol. 2. Lipkowitz, K. B., and Boyd, D. B. (Eds.). VCH: New York; 81–97 (1991).

[16] Press, W. H., Flannery, B. P., Teukolsky, S. A., and Vetterling, W. T. *Numerical Recipes in C*. Cambridge University Press: Cambridge 1988.

[17] Schlick, T. Optimization Methods in Computational Chemistry. In: *Reviews in Computational Chemistry*, Vol. 3. Lipkowitz, K. B., and Boyd, D. B. (Eds.). VCH: New York; 1–71 (1992).

[18] Eliel, E. L., Allinger, N. L., Angyal, S. J., and Morrison, G. A. *Conformational Analysis*. Wiley-Interscience: New York 1965.

[19] Pople, J. A. *Acc. Chem. Res.* **3**, 217 (1970).

[20] Hehre, W. J., Radom, L., Schleyer, P. v. R., and Pople, J. A. *Ab Initio Molecular Orbital Theory*. Wiley-Interscience: New York 1986.

[21] Szabo, A., and Ostlund, N. S. *Modern Quantum Chemistry: Introduction to Advanced Electronic Structure Theory*. McGraw-Hill: New York 1985.

[22] Clark, T. *A Handbook of Computational Chemistry: A Practical Guide to Chemical Structure and Energy Calculations*. Wiley-Interscience: New York 1985.

[23] De Frees, D. J., Levi, B. A., Pollack, S. K., Hehre, W. J., Binkley, S. J., and Pople, J. A. *J. Am. Chem. Soc.* **101**, 4085–4089 (1979).

[24] Davidson, E. R., and Feller, D. *Chem. Rev.* **86**, 681–696 (1986).

[25] Feller, D., and Davidson, E. R. Basis Sets for Ab Initio Molecular Orbital Calculations and Intermolecular Interactions. In: *Reviews in Computational Chemistry*, Vol. 1. Lipkowitz, K. B., and Boyd, D. B. (Eds.). VCH: New York; 1–43 (1990).

[26] Boyd, D. B. Aspects of Molecular Modeling. In: *Reviews in Computational Chemistry*, Vol. 1. Lipkowitz, K. B., and Boyd, D. B. (Eds.). VCH: New York; 321–354 (1990).

[27] Kunz, R. W. *Molecular Modelling für Anwender*. Teubner: Stuttgart 1991.

[28] Dewar, M. J. S., Zoebisch, E. G., Healy, E. F., and Stewart, J. J. P. *J. Am. Chem. Soc.* **107**, 3902–3909 (1985).

[29] Stewart, J. J. P. Semiempirical Molecular Orbital Methods. In: *Reviews in Computational Chemistry*, Vol. 1. Lipkowitz, K. B., and Boyd, D. B. (Eds.). VCH: New York; 45–81 (1990).

[30] Stewart, J. J. P. *J. Comput. Chem.* **10**, 209–220 (1989).

[31] Stewart, J. J. P. *J. Comput. Chem.* **10**, 221–264 (1989)

[32] Warshel, A. and Levitt, M. *J. Mol. Biol.* **103**, 227–49 (1976).

[33] CHARMM, Harvard University, Cambridge, Massachusetts, USA

[34] QSite, Schrödinger Inc., Portland, Oregon, USA

[35] QuanteMM, Accelrys, USA.

[36] Bash, P.A., Field, M.J., Davenport, R.C., Petsko, G.A., Ringe, D. and Karplus, M. *Biochemistry* **30**, 5826–5832 (1991).

[37] Mulholland, A.J. and Richards, W.G. *Proteins* **27**, 9–25 (1997).

2.3
Conformational Analysis

Molecules are not rigid. The motional energy at room temperature is large enough to let all atoms in a molecule move permanently. That means that the absolute positions of atoms in a molecule, and of a molecule as a whole, are by no means fixed and that the relative location of substituents on a single bond may vary in the course of time. Therefore, each compound containing one or several single bonds is existing at each moment in many different so-called *rotamers* or *conformers*. The quantitative and qualitative composition of this mixture is permanently changing. Of course only the low-energy conformers are found to a large extent.

A transformation from one conformation to another is primarily related to changes in torsion angles about single bonds. Only minor changes of bond lengths and angles take place. The changes in molecular conformations can be regarded as movements on a multi-dimensional surface that describes the relationship between the potential energy and the geometry of a molecule. Each point on the potential energy surface represents the potential energy of a single conformation. Stable conformations of a molecule correspond to local minima on this energy surface. The relative population of a conformation depends on its statistical weight which is influenced not only by the potential energy but also by the entropy. As a consequence, the global minimum on the potential energy surface—the conformation which contains the lowest potential energy—does not necessarily correspond to the structure with the highest statistical weight (for a detailed description see [1]).

Well-known examples for multiple conformations of molecules are the staggered and eclipsed forms of ethane, the anti-*trans* and *gauche* forms of *n*-butane or the boat and chair forms of cyclohexane. The rotation about the C_{sp3}–C_{sp3} bond in the ethane molecule can be described by a sine-like curve potential function (Fig. 2.5). The energy minima, located at 60°, 180° and 300°, correspond to the staggered form, while the maxima, located at 120°, 240° and 360°, correspond to the eclipsed form of ethane. Because structures located at maxima on the potential energy function (or potential energy surface) are not viable normally, only the staggered form of ethane needs to be taken into account when physical or chemical properties are studied. This straightforward situation completely changes in the case of larger and more flexible molecules which exist at room temperature in several energetically accessible rotamers. For example, at room temperature approximately 70% of *n*-butane exist in the anti-*trans* form and 30% in the *gauche* form [2]. Thus, for a discussion of the physical behavior of this flexible aliphatic chain both the anti-*trans* and the *gauche* conformations have to be taken into account. The same is true for cyclic structures like cyclohexane, where the chair as well as the boat form must be regarded.

The biological activity of a drug molecule is supposed to depend on one single unique conformation hidden among all the low-energy conformations [3]. The search for this so-called bioactive conformation for sets of compounds is one of the major tasks in medicinal chemistry. Only the bioactive conformation can bind to the specific macromolecular environment at the active site of the receptor protein.

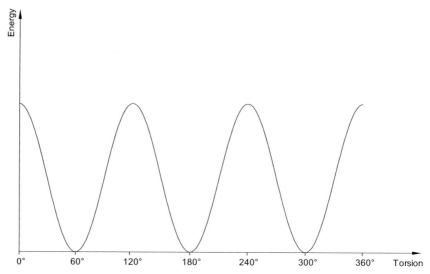

Fig. 2.5 Sine-like potential energy curve of ethane shown as function of the dihedral angle.

Based on the information of the active conformation one may be able to design new agents for a particular receptor system. It is widely accepted that the bioactive conformation is not necessarily identical with the lowest-energy conformation. However, on the other hand it cannot be a conformation that is so high in energy that it is excluded from the population of conformations in solution (for a discussion of this aspect see [4]). Thus, the identification of low-energy conformations is an important part of understanding the relationship between the structure and the biological activity of a molecule.

Experimental techniques such as NMR only provide information on one or a few conformations of a molecule. A complete and exclusive overview of the conformational potential of molecules can be gained by theoretical techniques. Correspondingly a variety of theoretical methods for conformational analysis has been developed. Many applications are reported in the literature [5–12]. The most general methods for conformational analysis are those that are able to identify all minima on the potential energy surface. However, as the number of minima increases dramatically with the number of rotatable bonds, an exhaustive detection of all minima becomes a difficult and time-consuming task.

The time required for a conformational analysis depends also directly on the type of method used for the calculation of the energy. Conformational energies can be calculated either using quantum mechanical or molecular mechanical methods. Because the quantum mechanical calculations are very time consuming, they cannot be applied to large or flexible molecules. For that reason most of the conformational search programs use molecular mechanics methods for the calculation of energies as a standard. Apart from systematic search procedures we also will deal in this chapter with the use of Monte Carlo and molecular dynamics techniques for conformational analyses.

2.3.1
Conformational Analysis Using Systematic Search Procedures

The systematic search [6,7,13] is perhaps the most natural of all different conformational analysis methods. It is performed by varying systematically each of the torsion
angles of a molecule in order to generate all possible conformations. If the angle
increment is appropriately small the procedure yields a complete image of the conformational space of any molecule.

The step size which is normally used in a systematic search is 30°. That means,
during a full rotation of 360°, 12 conformations are generated. In close neighbor-
hood to the optimal value a smaller step size down to 5° may be necessary in order
to determine the minimum position of a conformation exactly. The number of generated conformations depends on the step size, but also on the number of rotatable
bonds. If n is the number of rotatable bonds, then the number of conformations
increases with the nth power:

Number of conformations = $(360/\text{step size})^n$

If for example a systematic conformational search is performed for a molecule with
six rotatable bonds and a step size of 30° is employed, the number of generated conformers amounts to 12^6 or 2 985 984 structures. This huge amount of data cannot
be handled; it therefore has to be reduced.

The first step in data reduction is a van der Waals screening or ‚"bump check". It
is performed before the potential energy of the conformations will be exactly calculated. The screening procedure excludes all conformations where a van der Waals
volume overlap of atoms not directly bound to each other is detected. The mathematical criterion for determining the validity of a conformation in this respect simply is
the sum of the van der Waals radii of non-bonded atoms. The hardness of van der
Waals spheres can be varied by specification of a so-called van der Waals factor. This
multiplication constant controls the interpenetrability of atoms. A reduction of the
van der Waals factor results in softening of contacts between non-bonded atoms,
thereby increasing the number of valid conformations.

For the conformers remaining after the bump check the potential energy is calculated using a molecular mechanics method. In general the conformational energy is
calculated neglecting electrostatic interactions, i.e. charges are not taken into
account and the conformational analysis is performed in vacuo. The reasons for this
procedure have been discussed in section 2.2.4. If the inclusion of electrostatic interactions into a conformational analysis is justified for a special case then the whole
process becomes much more complex. Good-quality atomic charges are sensitive to
the discrete spatial environment of the atoms and not only depend on the connectivity. Therefore, atomic charges which have been calculated for the initial conformation must be constantly updated after each modification of a torsion angle. In addition it would be necessary to mimick the effect of a solvent, which tones down the
strong electrostatic interactions built up between charges in vacuo. Obviously this
procedure would require a large amount of additional computer time, even for a

small molecule. And what is even more noteworthy, the increase in complexity of the system does not produce a deeper insight into the conformational behavior of a molecule in solution, besides the fact that intramolecular interactions are diminished. The same result is obtained when charges are not considered and the analysis is performed in vacuo. Besides, in the active site of a receptor or enzyme the intramolecular contacts in ligands are also of minor importance.

When the conformational energies have been calculated for all conformers which survived the bump check another possibility to reduce the number of conformations is the use of an energy window. The underlying idea for applying an energy window is based on the fact that conformations containing much more energy than those close to the minimum are found in the conformer population only to a neglectable quantity, i.e. in our context it may be assumed that they do not have any importance for the biological activity of a particular molecule. The value for this energy window depends on the size of the studied molecule as well as on the applied force field. It may vary between 5 and 15 kcal mol^{-1} [11–15].

The resulting conformations, which have passed all filter methods should represent a complete ensemble of energetically accessible conformations for a particular molecule. However, in many cases the number still may be too large to allow a reasonable treatment. Many of the remaining conformations are very strongly related, because they only differ for example in a single rotor step. Obviously these can be combined to a common family with pronounced similarity. The description of the conformational properties of a molecule does not lack comprehensiveness if we only take the minimum conformer of each conformational family into further consideration. Several methods have been developed to execute the classification into conformational families [15–17]. The parameters used for this purpose are the torsion angles. The known classification methods differ in the procedure to associate the conformers to individual families. Another possibility to evaluate the large amount of data accumulated in course of a systematic conformational search is the application of statistical techniques like cluster or factor analysis. For a detailed discussion of these methods see [18].

The course of a systematic conformational analysis shall be demonstrated on a study performed in our group with two H_2-antihistaminic agents, tiotidine and ICI127032 (Fig. 2.6) [19]. It was performed using the SEARCH module within the molecular modeling package SYBYL [16].

As rotational increment a step size of 15° was chosen. Due to symmetry the methyl group of the cyanoguanidine system was only rotated in steps of 30° between 0° and 120°. The theoretical number of conformations, 3.98×10^7, was reduced using the van der Waals screening to 4.6×10^6, i.e. roughly 10% of the initial number still is valid after the bump check. The application of an energy window of 15 kcal mol^{-1} leads to a further reduction of 90%. Some 453 393 conformations were stored. Even this number cannot be handled in a reasonable way. Therefore, in a next step the conformations left were classified into families using the program IXGROS [17] which has been developed in our group. This finally yields 227 unique families which are represented by their respective minimum energy conformations. Although the reduction from 4.0×10^7 down to 227 conformations is very impress-

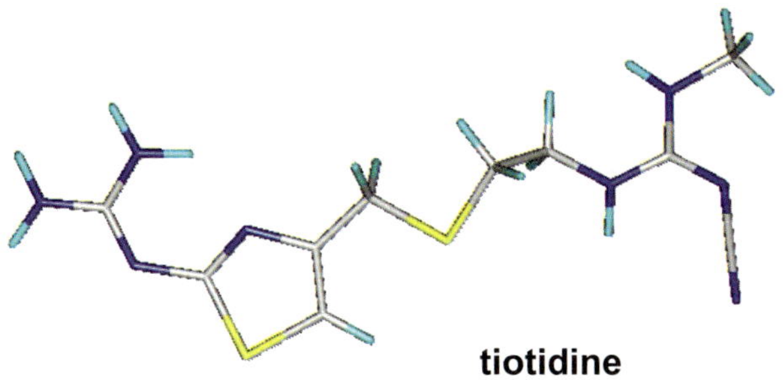

tiotidine

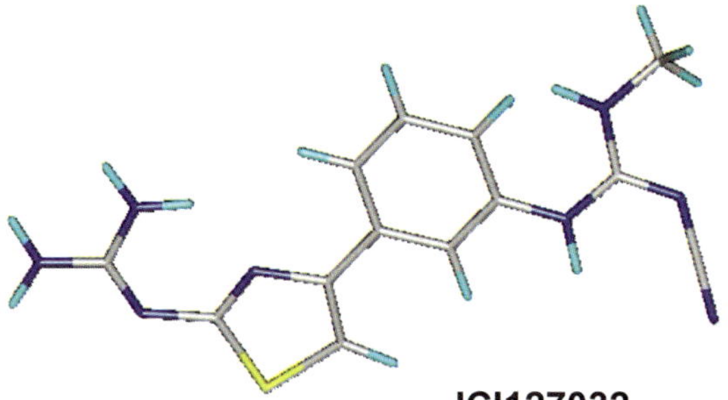

ICI127032

Fig. 2.6 Molecular formulas of the histamine H_2 receptor antagonists tiotidine and ICI127032.

ive, one has to submit that even the rather small number left is too large. There is no chance to decide which of the 227 conformers is the bioactive one, but this and only this is the question of interest. At this point a solution cannot be found if there do not exist rigid or at least semi-rigid congeners which in addition must be biologically active. It also must be proven that they bind to the same receptor site in an analogous mechanism. That is, as a rule, for finding the bioactive conformation of a flexible molecule, potent and more rigid compounds of the same series are needed. In the case of the H_2 antagonists the rigid and potent representative is ICI127032. After consideration of the small number of low-energy conformations of the rigid matrix and repeated use of IXGROS, eight unique families survived the procedure. These remaining conformations could be used successfully to determine the biologically active conformation of tiotidine (Fig. 2.7).

As we have discussed it is of advantage to include rigid molecules in a conformational search for a set of flexible congeners. The rigid and biologically potent derivatives are used as a matrix for all other members of the series. Marshall and colleagues [7] have extended this procedure by also including inactive rigid representatives. By doing this the conformational space can be further restricted and by the same token the time necessary for the search is reduced by orders of magnitude. This technique has become known as "Active analogue approach".

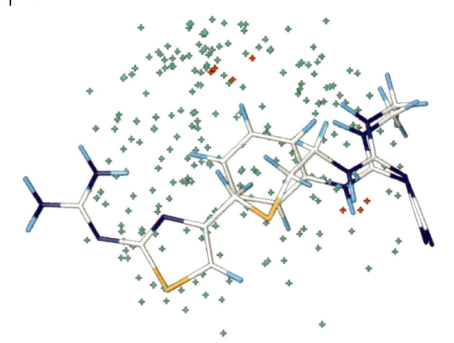

Fig. 2.7 Representation of the results of the conformational analysis of tiotidine and ICI127032 (both displayed in a possible minimum energy conformation). The local minimum conformation representing the different conformational families are displayed by stars symbolizing the center of the cyanoguanidine end group of tiotidine and ICI127032. The resulting conformations of tiotidine are indicated by green stars, while the red-coloured stars mark the conformations derived for ICI127032. (The calculations have been performed using the SEARCH module within SYBYL 6.1 [16] and IXGROS [17]).

2.3.2
Conformational Analysis Using Monte Carlo Methods

A completely different path for searching conformational space is realized in the Monte Carlo or random search. Random search techniques are of a statistical nature [20]. At each stage of a Monte Carlo search the actual conformation is modified randomly in order to obtain a new one.

A random search starts with an optimized structure. At each iteration in the procedure, new torsion angles [11] or new cartesian coordinates [8,9] are assigned randomly. The resulting conformation is minimized using molecular mechanics and the randomization process is repeated. The minimized conformation is then compared with the previously generated structures and is only stored if it is unique. The random methods potentially cover all regions of conformational space, but this only is true if the process is allowed to run for a sufficiently long time. This may last extremely long because the probability to detect a new and unique conformation decreases dramatically depending on the growing number of conformers already discovered. However, even if the computation has been running very long, one cannot be certain that the conformational space has been completely covered. It is very important therefore to establish a means for testing the completeness of the analysis. This can be done efficiently by performing several runs in a parallel mode, each one starting with a different initial conformation. If the results are identical or nearly identical, then completeness can be assumed. Another measure of completeness is based on the recovery rate for each low-energy conformation, because the probabilistic process must reproduce it many times.

The main advantage of random search methods is that, in principle, molecules of any size can be successfully treated. In practice, however, highly flexible molecules often do not give converging results, because the volume of the respective conformational space is too large. Other useful applications for Monte Carlo search methods include investigations on cyclic systems, because ring systems in general are difficult to treat in systematic searches. The effectiveness of random search procedures shall be demonstrated on a practical example. Cycloheptadecane was studied using a variety of different methods including a random search method [12]. The combined results of the various procedures yielded a total amount of 262 different minimum conformations. None of the employed techniques succeeded in finding all 262 conformers, but one of the random search analyses nevertheless was able to detect 260 of them. It is therefore safe to comment that random search techniques are very suitable for conformational analyses of many types of molecules, but may require a large amount of computer time to ensure complete coverage of conformational space.

Another sampling technique widely applied to the problem of improved conformational searching is known as poling [21]. Conformational variation is promoted through the addition of a so-called poling function to a standard molecular mechanics force field. The poling function has the effect of changing the energy surface being minimized to penalize conformational space around any previously accepted conformers. As a consequence the method both increases conformational variation and eliminates redundancy within the limits imposed by the function. Poling has been applied within the CATALYST program [22] in order to search large databases of molecules.

2.3.3
Conformational Analysis Using Molecular Dynamics

The systematic conformational search procedure is a valuable tool to determine the large number of minima on the potential energy surface associated with a flexible molecule. In principle, the generation of all allowed conformations can be realized and there is a high probability for the completeness of the conformational search. However, there are clear limitations in the applicability of this method. The multiminima problem can only be solved for rather small molecules with a limited number of rotatable bonds.

As already mentioned in section 2.3.1 the systematic conformational search of a molecule with six rotatable bonds leads to serious problems in data handling due to the large number of generated conformers. Therefore the investigation of flexible molecules—like for example arachidonic acid (Fig. 2.8), which contains 15 rotors—is practically impossible. Even after applying several methods of data reduction the systematic conformational search for this molecule yielded 500 000 different conformations. The procedure was stopped automatically by the program due to data overflow, although the conformational space was not completely sampled at this point.

However, conformational analysis of the same molecule by a random search procedure will also be unreasonable because of the required computer time. For example, the investigation of cycloheptadecane—which is a more restricted molecule—used about 94 days of computer time on a Micro-Vax II computer [12].

Another rather difficult subject in this context is presented when saturated or partially saturated ring systems are to be treated in a systematic conformational analysis. In the course of the systematic process, bonds have to be broken in order to produce new attainable ring conformations. Efficiency and reliability of this procedure have been subject of several reviews [12, 14].

A very common strategy to overcome these problems is the use of molecular dynamics simulations for exploring conformational space. The aim of this approach is to reproduce the time-dependent motional behavior of a molecule. Molecular dynamics are based on molecular mechanics. It is assumed that the atoms in the molecule interact with each other according to the rules of the employed force field (as already described in section 2.2.1). At regular time intervals the classical equation of motion represented by Newton's second law is solved:

$$F_i\,(t) = m_i\,a_i\,(t) \tag{8}$$

where F_i is the force on atom i at time t, m_i is the mass of atom i, and a_i is the acceleration of atom i at time t. The gradient of the potential energy function is used to calculate the forces on the atoms while the initial velocities on the atoms are generated randomly at the beginning of the dynamics run. Based on the initial atom coordinates of the system, new positions and velocities on the atoms can be calculated at time t and the atoms will be moved to these new positions. As a result of this a new conformation is created. The cycle will then be repeated for a predefined

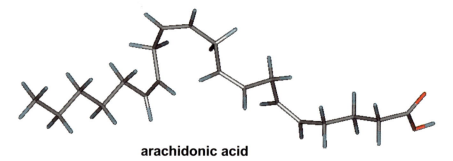

arachidonic acid

Fig. 2.8   One energetically permitted conformation of arachidonic acid.

number of time steps. The collection of energetically accessible conformations produced by this procedure is called an *ensemble*.

The application of Newton's equations of motion is uniform in all different available molecular dynamics approaches, but they differ in the employed integration algorithms. Very common methods for integrating the equations of motion are the Verlet integrator [23] and algorithms like Beeman [24] and the leap-frog scheme [25] which are simple modifications of the Verlet algorithm. In the framework of this book a more extended discussion of the molecular dynamics theory is not intended but the interested reader is urged to study more detailed reviews on this subject [26–29].

Before employing molecular dynamics simulations for conformational analysis the reader's attention should be drawn to some special features of this method. Unlike the conservative geometry optimization procedures, molecular dynamics is able to overcome energy barriers between different conformations. Therefore it should be possible to find local minima other than the nearest in the potential energy surface. However, if the energy barrier is high or the number of degrees of freedom in the molecule is very large, then some of the existing conformers of the investigated system possibly are not reached. In view of the huge conformational space the completeness of the conformational search during the chosen simulation time is difficult to ensure.

To enhance conformational sampling a widely used tactics in molecular dynamics is to apply an elevated temperature to the simulation [29]. At high temperature the molecule is able to overcome even large energy barriers that may exist between some conformations and therefore the chance for completeness of a conformational search increases. It is self evident that the choice of a particular simulation temperature and simulation time depends closely on the molecule of interest.

One recent and comprehensive investigation can be used to demonstrate the dependence of conformational flexibility on the simulation temperature. The data and additional material were made available by courtesy of F. S. Jorgensen, Royal Danish School of Pharmacy, Copenhagen (Denmark). A molecular dynamics simulation has been performed on the experimentally well-studied cyclohexane molecule using different start conformations and different simulation temperatures[*] (Fig. 2.9).

At 400 K the twist form of cyclohexane ($T_1 = 0$) which has been used as initial conformation, oscillates between different twist forms while at 600 K the molecule contains sufficient kinetic energy to convert to one of the chair conformations ($T_1 = 300$). Further increase of the temperature up to 1000 K yielded both chair as well as twist conformations ($T_1 = 300 \rightarrow 60$) and several chair–chair interconversions can be observed. After 800 ps one of the chair conformations ($T_1 = 60$) exists almost exclusively. In a second study, three methyl-substituted cyclohexanes (1,1-dimethyl-cyclohexane, 1,1,3,3-tetramethylcyclohexane and 1,1,4,4-tetramethylcyclohexane) were subjected to molecular dynamics simulations at various temperatures. The

---

[*] Sybyl (version 6.0.3) from Tripos Associates Inc., St. Louis, U.S.A. Energy minimizations: Tripos force field, PM3 partial charges, dielectric constant $Å = 20$ and a convergence criterion of 0.005 kcal mol^{-1}Å^{-1}. MD simulations: 1000 ps at various temperatures with conservation of total energy, one conformation sampled per picosecond.

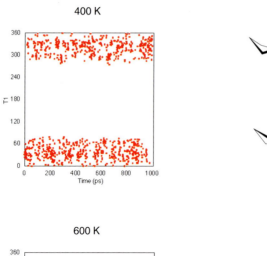

400 K

600 K

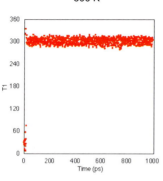

1000 K

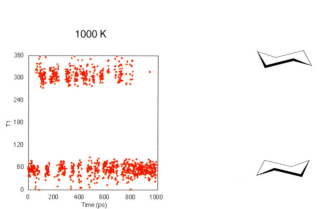

Fig. 2.9 Variation of torsion angle T_1 (torsion angle $T_1 = C_1-C_2-C_3-C_4$) of cyclohexane for different simulation temperatures. At 400 K the molecule oscillates between different flexible twisted boat forms reflected by an extensive fluctuation of the observed torsion angle.

Increasing the temperature to 600 K leads to one of the possible stable minima corresponding to one chair conformation. The dynamic simulation at 1000 K yields both chair conformations as well as the already observed twist and boat conformations.

observed chair–chair interconversions at the corresponding temperatures have been compared with experimentally determined energy barriers of ring inversion [30] (Table 2.2). As a result of the comparison it can be concluded that the molecular dynamics simulations are able to reflect the relative magnitude of the experimentally determined ring inversion barriers. This example of high temperature molecular dynamics clearly indicates the necessity to verify if the chosen simulation temperature is high enough to prevent the system from getting stuck in one particular region of conformational space.

Tab. 2.2 Data on the existence of the two possible chair conformations (chair and chair') of three methyl-substituted cyclohexanes at different simulation temperatures. The data are compared with the corresponding experimentally determined ring inversion barriers

	Temperature				
Molecular form	600 K	800 K	1000 K	1200 K	ΔG (kcal/mol^{-1})
	Chair	Chair + Chair'	Chair + Chair'	Chair + Chair'	9.6
	Chair	Chair	Chair + Chair'	Chair + Chair'	10.6
	Chair	Chair	Chair	Chair + Chair'	11.7

In the application of molecular dynamics to search conformational space it is a common strategy to select conformations at regular time intervals and minimize them to the associated local minimum. This procedure has been used in several conformational analysis studies on small molecules, including ring systems [14,31]. A very impressive example in this context is the conformational analysis of the polyhydroxy analog of the sesquiterpene lacton tharpsigargin (Fig. 2.10). This study was also performed in the laboratory of F. S. Jorgensen.

The polyhydroxy derivative has been studied in molecular dynamics simulations at 1200 K in order to gain insight into the conformational behavior of the ring system[*]. The seven-membered ring adopted several different conformations during the simulation and a considerable number of ring interconversions took place. This clearly demonstrates an extensive exploration of the conformational space.

Each of the sampled conformations has been energy-minimized subsequently and compared exclusively with respect to the conformation of the seven-membered ring. All conformations with a root mean square (rms) value below 0.1 Å were con-

[*] Sybyl (version 6.0.3) from Tripos Associates Inc., St. Louis, U.S.A. Energy minimizations: Tripos force field, PM3 partial charges, dielectric constant Å = 20 and a convergence criterion of 0.005 kcal mol^{-1} Å^{-1}. MD simulations: 1000 ps at 1200 K with conservation of total energy, one conformation sampled per picosecond.

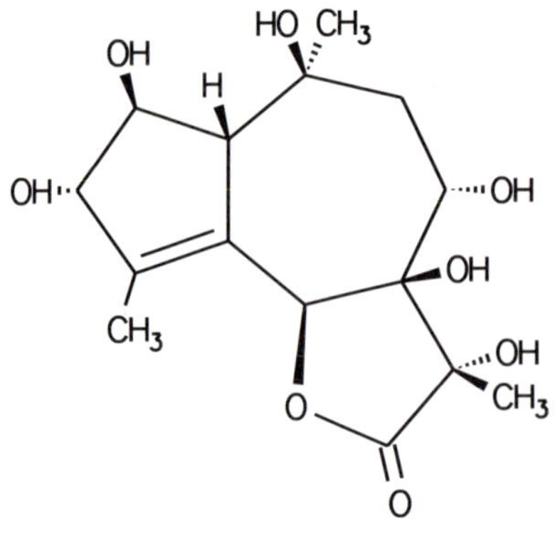

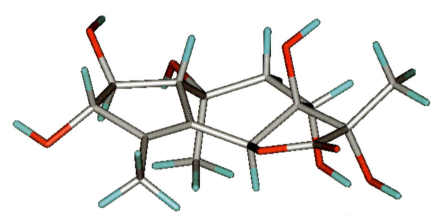

Fig. 2.10 Molecular formula of the polyhydroxy analog of tharpsigargin.

sidered to be identical. The procedure yielded five different low-energy conformations. Fortunately NMR data [32] of tharpsigargin agree with one of the theoretically found conformations of the tricyclic ring system. This is shown in Fig. 2.11.

In some cases, however, it is not sufficient to minimize the sampled conformations in order to reach the final minimum conformation. The intention of the high-temperature dynamics

simulation is to provide the molecule with enough kinetic energy to cross energy barriers between different conformations. However, during the simulation the molecule can occupy extremely distorted geometries which sometimes cannot be relaxed by a simple minimization procedure.

If this occurs it is recommended to perform a high-temperature annealed molecular dynamics simulation [33]. Using this approach all sampled conformations of the high-temperature simulation will be subsequently optimized and then reshaken at a lower temperature, e.g. 300 K, in order to remove the internal strain of the molecule. The final reoptimization leads to conformations of lower energy when compared

Fig. 2.11 One of the theoretically determined conformations of the polyhydroxy analog of tharpsigargin. The ring conformation is in accordance with results obtained by NMR spectroscopy.

with the results of a high-temperature simulation which is followed by a simple geometry optimization.

An additional modification of this high-temperature annealed molecular dynamics simulation is the so-called simulated annealing method [34]. In this technique the system is cooled down at regular time intervals by decreasing the simulation temperature. As the temperature approaches 0 K the molecule is trapped in the nearest local minimum conformation. The received geometry at the end of the annealing cycle is saved and subsequently used as starting point for further simulations at high temperature. In order to obtain a set of low-energy conformations the cycle will be repeated several times. As the resulting structures should already be close to a minimum it is not absolutely necessary subsequently to minimize the structure. The application of this method has been subject of several studies [35, 36]. Further information may be found in these references.

In conclusion, it may be stated that molecular dynamics simulations represent an additional and very valuable tool that can be used to sample the conformational space, especially when other conformational search methods have been unsuccessful. The user should be careful when selecting the appropriate method and in setting the simulation conditions in order to ensure the completeness of the conformational search and the validity of the results. It should also be kept in mind that each approach has its strengths and its weaknesses and therefore, wherever possible, experimentally derived data should serve as verification.

2.3.4
Which is the Method of Choice?

With such a variety of methods for sampling conformational space, it can be difficult to make the right choice. Each method has its own strengths and weaknesses. Systematic searches are subject to the effects of conformational explosion, and they cannot normally be applied to cyclic molecules. Random search methods can require long runs to ensure that the conformational space has been completely sampled. In addition, duplicates have to be removed from the output.

Conformational search methods are typically validated against standard benchmark data sets. In one typical approach, several parameters are used to test both the number of conformers as well as their energies. An alternative approach is to compare generated conformations against experimentally derived X-ray structures taken from the Cambridge Crystallographic Database [37]. Sadowski et al. found that the CORINA program, which is used to convert 2D into 3D structures, reproduced the correct conformation for nearly half of their data set of 639 structures [38].

Since the number of high-resolution X-ray protein-ligand complexes included in the Protein Database [39] is increasing rapidly, a further approach for testing conformational search methods now exists. It is possible to detect whether experimentally determined bioactive conformations are present in various generated conformational ensembles. A comparison of different sampling techniques has recently been published [40]. In this study, it was investigated to which extent the results of several conformational searching tools agree with the experimental results. The authors applied systematic and random search methods to a set of 32 diverse small molecules for which the structure of the receptor-ligand complex had also been obtained by X-ray crystallography [40]. The comparison showed that the Low-Mode Conformational Search method of the MacroModel program [41] performed better than other algorithms. Reducing the intramolecular electrostatic interactions, either by including a solvation model or by neglecting atomic charges, had a benefical effect on the likelihood of finding the bioactive conformation. It proved difficult to retrieve structures having more than eight rotatable bonds for all methods. According to these results a variety of ligands do not bind in a minimum energy conformation to the protein. Similar observations have been reported by other authors [42, 43].

As a conclusion it may be stated that presently a variety of methods exist which may be used to sample the conformational space. The user should be careful when selecting the appropriate method and in setting the simulation conditions, in order to ensure the completeness of the conformational search and the validity of the results. It should be kept in mind that each approach has its strengths and its weaknesses and, therefore, experimentally derived data should serve as verification wherever possible.

References

[1] Scheraga, H. A. *Chem. Rev.* **71**, 195–217 (1971).

[2] Rademacher, P. *Strukturen organischer Moleküle.* VCH: Weinheim 1987, p. 139.

[3] Ghose, A. K., Crippen, G. M., Revankar, G. R., Smee, D. F., McKernan, P. A., and Robins, R. K. *J. Med. Chem.* **32**, 746–756 (1989).

[4] Jörgensen, W. L. *Science* **254**, 954–963 (1991).

[5] Howard, A. E., and Kollman, P. A. *J. Med. Chem.* **31**, 1669–1675 (1988).

[6] Lipton, M., and Still, W. C. *J. Comput. Chem.* **9**, 343–355 (1988).

[7] Dammkoehler, R. A., Karasek, S. F., Shands, E. F. B., and Marshall, G. R. *J. Comput.-Aided Mol. Design* **3**, 3–21 (1989).

[8] Saunders, M. *J. Am. Chem. Soc.* **109**, 3150–3152 (1987).

[9] Saunders, M. *J. Comput. Chem.* **10**, 203–208 (1989).

[10] Ferguson, D. M, and Raber, D. J. *J. Am. Chem. Soc.* **111**, 4371–4378 (1989).

[11] Chang, G., Guida, W. C., and Still, W. C. *J. Am. Chem. Soc.* **111**, 4379–4386 (1989).

[12] Saunders, M., Houk, K. N., Wu, Y.-D., Still, W. C., Lipton, M., Chang, G., and Guida, W. C. *J. Am. Chem. Soc.* **112**, 1419–1427 (1990).

[13] Ghose, A. K., Jaeger, E. P., Kowalczyk, P. J., Peterson, M. L., and Treasurywala, A. M. *J. Comput. Chem.* **14**, 1050–1065 (1993).

[14] Böhm, H.-J., Klebe, G., Lorenz, T., Mietzner, T., and Siggel, L. *J. Comput. Chem* **11**, 1021–1028 (1990).

[15] Taylor, R., Mullier, G. W., and Sexton, G. J. *J. Mol. Graphics* **10**, 152–160 (1992).

[16] SYBYL Theory Manual, Tripos Associates, St. Louis, Missouri, USA.

[17] Sippl, W. *Theoretische Untersuchungen zum Bindungsverhalten von Histamin H₂- und H₃-Rezeptor Liganden*, Ph. D. thesis, Heinrich-Heine-Universität Düsseldorf, Germany 1997.

[18] Shenkin, P. S., and McDonald, D. Q. *J. Comput. Chem.* **15**, 899–916 (1994).

[19] Höltje, H.-D., and Batzenschlager, A. *J. Comput.-Aided Mol. Design* **4**, 391–402 (1990).

[20] Metropolis, N., Rosenbluth, A. W., Rosenbluth, M. N., Teller, A. H., and Teller, E. *J. Chem. Phys.* **32**, 1087–1092 (1953).

[21] Smelie, A., Kahn, S. D., and Teig, S. L. *J. Chem. Inf. Comput. Sci.*, **2**, 295–304 (1995).

[22] CATALYST, Accelrys Inc., San Diego, California, USA.

[23] Verlet, L. *Phys. Rev.* **159**, 98–103 (1967).

[24] Beeman, D. *J. Comp. Phys.* **20**, 130 (1976).

[25] Hockney, R. W., and Eastwood, J. W. *Computer Simulation Using Particles.* McGraw-Hill: New York 1981.

[26] van Gunsteren, W. F., and Berendsen, H. J. C. *Angew. Chemie* **102**, 1020–1055 (1990).

[27] Lybrand, T. P. Computer Simulation of Biomolecular Systems Using Molecular Dynamics and Free Energy Perturbation Methods. In: *Reviews in Computational Chemistry*, Vol. 1. Lipkowitz, K. B., and Boyd, D. B. (Eds.). VCH: New York; 295–320 (1990).

[28] McCammon, J. A., and Harvey, S. C. *Dynamics of Protein and Nucleic Acids.* Cambridge University Press: Cambridge 1987.

[29] Leach, R. A. A Survey of Methods for Searching the Conformational Space of Small and Medium-Sized Molecules. In: *Reviews in Computational Chemistry*, Vol. 2. Lipkowitz, K. B., and Boyd, D. B. (Eds.). VCH: New York; 1–47 (1991).

[30] Friebolin, H., Schmid, H. G., Kabuß, S., and Faißt, W. *Org. Magn. Reson.* **1**, 147–162 (1969).

[31] Kawai, T., Tomioka, N., Ichinose, Takeda, M., and Itai, A. *Chem. Pharm. Bull.* **42**, 1315–1321 (1994).

[32] Christensen, S. B., and Schaumburg, K. *J. Org. Chem.* **48**, 396–399 (1983).

[33] Auffinger, P., and Wipff, G. *J. Comput. Chem.* **11**, 19–31 (1990).

[34] Kirkpatrick, S., Gelatt, C. D., Vecchi, M. P. *Science* **220**, 671–680 (1983).

[35] Salvino, J. M., Seoane, P. R., and Dolle, R. E. *J. Comput. Chem.* **14**, 438–444 (1993).

[36] Laughton, C. A. *Protein Eng.* **7**, 235–241 (1994).

[37] Cambridge Structural Database, Dr. Olga Kennard, F.R.S., Cambridge Crystallographic Data Centre, 12 Union Road, Cambridge CB2, 1EZ, U.K.

[38] Sadowski, J., Gasteiger, J. and Klebe, G. *J. Chem. Inf. Comput. Sci.* **34**, 1000–1012 (1994).

[39] Bernstein, F., Koetzle, T.F., Williams, G.J.B., Meyer Jr, E.F., Brice, M.D., Rodgers, J.R., Kennard, O., Shimanouchi, T. and Tasumi, M.J. *J. Mol. Biol.* **112**, 535–542 (1977).

[40] Boström, J. *J. Comput.-Aided Mol. Design* **15**, 1137–1152 (2001).

[41] MacroModel V7.0: Mohamadi, F., Richards, N.G.J., Guida, W.C., Liskamp, R., Lipton, M., Caufield, C., Chang, G., Hendrikson, T. and Still, W.C. *J. Comput. Chem.* **11**, 440–456, (1990).

[42] Vieth, M *J. Comput.-Aided Mol. Design* **12**, 563–572 (1998).

[43] Boström, J., Norrby, P.-O. and Liljefors, T. *J. Comput.-Aided Mol. Design* **12**, 383–396 (1998).

2.4
Determination of Molecular Interaction Potentials

The initial step in the formation of a complex like, for example, a drug–receptor complex is a recognition event. The receptor has to recognize whether an approaching molecule possesses the properties necessary for specific and tight binding. This recognition process occurs at rather large distances and precedes the formation of the final interaction complex. The 3D electrostatic field surrounding each molecule therefore plays a crucial role in recognition. Other molecular characteristics like polarizability or hydrophobicity come into play when the distance between the interacting surfaces gradually decreases. It is therefore easy to realize that molecular fields which can be determined by systematic calculation and sampling of interaction energies between the molecules under study using different chemical probes represent data sets of high value for the understanding of intermolecular interaction at any level of complexity of the molecular ensemble of interest.

In the following sections the methods for calculation and analysis of these molecular properties will be described and evaluated.

2.4.1
Molecular Electrostatic Potentials (MEPs)

Knowledge of the molecular electrostatic potential (MEP) is critically important when molecular interactions and chemical reactions are to be studied. If molecules approach each other, the initial contact arises from long-range electrostatic forces. In principle, interaction forces can be separated into three components: electrostatic, inductive and dispersive. The first type of interaction appears between polar molecules which carry a charge or possess a permanent dipole moment. The second type is found when a polar molecule interacts with a non-polar molecule. The dipole of the polar molecule then produces an electric field which changes the distribution of the electrons in the non-polar molecule, thereby inducing a dipole moment. Thirdly, even if both molecules are non-polar and hydrophobic entities, the permanent fluctuations in the electron distribution of one molecule can induce a temporary molecular dipole moment in a neighboring molecule. This type of interaction is called *dispersion*. Dispersion forces are weak and fall off rapidly with increasing distance between the interacting molecules (see section 2.2.1). However, they constitute the main part of attraction between neutral non-polar molecules. (The dispersion forces are also called London forces.)

The electrostatic interaction can be either attractive or repulsive; an electropositive portion of an approaching molecule will seek to dock with an electronegative region, while similarly charged portions will repel each other. The non-covalent interaction obviously is especially large between charged regions of molecules. Due to charges—but also due to permanent dipole moments present in a molecule—a 3D electrostatic field is generated in the surrounding environment. Therefore at moderate distances from polar or even neutral molecules, a significant molecular electrostatic potential exists. This can be represented as interaction energy between the molecular electron distribution and a positive point charge which is located in a 3D grid at any point in space surrounding the molecule. For the determination of the molecular electrostatic potential an accurate treatment of the electronic properties of the molecules is required. Therefore, methods for the calculation of molecular charge densities become priority.

2.4.1.1 Methods for Calculating Atomic Point Charges

The electronic properties of molecules are defined through the electron distributions around the positively charged nuclei. Detailed information about the electron distribution can be either obtained via experimental results, e.g. X-ray diffraction studies, or by calculations using quantum mechanical methods. However, with respect to the computational procedure corresponding results provide only a probability distribution of the charge density throughout three-dimensional space. For the purpose of interaction energy calculations mostly point charges located at the center of the atom positions are needed. Without doubt this produces a very simplified picture of the molecular electron distribution. To achieve the transformation the electron density needs to be converted into so-called partial or point charges. This can be done by contracting the charge onto the atomic centers. Thus, the picture of a molecule consisting of atoms carrying the partial or point charges has emerged. The definition of these empirical partial charges bears some arbitrariness because the molecular electron distribution must be assigned to individual atom centers. Or to put it in a different way, a molecular characteristic is scaled down to an atomic property. Partial charges are not observable, so the method of assigning point charges is only relevant and scientifically sound when it can be used to correlate or predict physical or chemical properties of molecules. On the other hand, as stated before, the electrostatic part of the overall intermolecular interaction energy is very prominent and therefore most of the commonly used molecular mechanics programs include a corresponding energy term which is dependent on atomic partial charges. The application of these methods allows the rapid computation of electrostatic energies, even for macromolecules with more than a few hundred atoms. For that reason a variety of different techniques for the calculation of atomic partial charges has been developed (for a review, see [1]).

In principle it must be distinguished between two methodologically absolutely different approaches:

1. Topological procedures [2–6] such as the Gasteiger–Hückel method [2].
2. Procedures which calculate atomic charges from the quantum chemical wave functions like the population analysis [7] or the potential-derived charge calculation methods [8–11].

Topological Charges

The topological methods are based mainly on the electronegativity of the different atom types. To allocate atomic charges to directly bonded atoms in a reasonable way, appropriate rules are used which combine the atomic electronegativities with experimental structural informations on the bonds linking the atoms of interest. The topological methods do not need information about the molecular geometry or conformational status of a molecule. Only the connectivity matrix of the atoms is included in the calculation. The original method proposed by Del Re [3] exclusively for saturated molecules was extended to conjugated systems by Berthod and Pullman [4]. Both methods still are implemented in some modeling programs. A newer approach, which gives more realistic results in comparison with experimental data is the Gasteiger-Hückel method. It is a combination of the Gasteiger–Marsili method [2] for the calculation of the σ component of the atomic charge and the old Hückel theory [12]. The Hückel theory allows to calculate the π component of the atomic charge in a fast and fairly efficient way. Naturally the total charge is the sum of σ and π elements. Formal charges on atoms included in π systems are assumed to be delocalized over the whole π system. For this reason, Hückel charges are calculated first and the Gasteiger charge calculation is performed subsequently. The big advantage of the topological procedures is that they are computationally fast and in many cases compare quite well with experimentally observable properties. The big danger is that one cannot trust the results without validation for a particular group of molecules. Very often the validation procedure simply is omitted. Of course this renders the corresponding study useless.

Topological methods often are implemented into commercial software packages as standard tools for charge calculation.

Quantum Chemical Methods
All other methods for the calculation of atomic partial charges are based on the quantum mechanical computation of wavefunctions. Wavefunctions either can be obtained using semiempirical or ab initio methods depending on the requested accuracy of the wavefunction and also on the available computational resources. Charge densities can be obtained from wavefunctions using different procedures. The oldest and most widely used is the Mulliken population analysis [7], which is implemented as standard method in various quantum mechanical programs [13–15]. The population analysis takes the electron density derived from the wavefunction and partitions it between the atoms on the basis of the occupancy of each atomic orbital. Although widely used, it has long been recognized in the literature that the results of the Mulliken method depends strongly on the basis sets applied. It often gives unrealistic results [16,17] (see also Table 2.3). An improved technique that eliminates most of the problems associated with the Mulliken procedure is the natural population analysis [18], but it is effective on ab initio wavefunctions only.

A second, much more recently developed, technique yielding atomic charges from quantum mechanically calculated wavefunctions is the method of deriving charges by fitting the molecular electrostatic potential (also called electrostatic potential (ESP) fit method) [7–11]. The charge density is a well-defined function [19]. It contains important and detailed information about the molecule because all electrons contribute in some way to the distribution of the electronic charge in space. It also is experimentally accessible [20] from X-ray diffraction. However, this technique is extremely demanding as far as costs and time consumption are concerned and cannot be used as a standard procedure. A set of atomic charges able to reproduce

Table 2.3 Comparison of experimentally derived and theoretically calculated dipole moments. The theoretical dipole moments were calculated using several procedures: the Gasteiger–Hückel method was chosen as an example for simple topological methods; on the quantum mechanical level the dipole moments were calculated directly from the wavefunction (SCF) as well as using the Mulliken and potential-derived point charges (ESP)

Molecule	Experimental (gas phase)	Gasteiger-Hückel	AM1			PM3			STO-3G			3-21G*			6-31G**		
		SCF	SCF	Mulliken	ESP	SCF	Mulliken	ESP	SCF	Mulliken	ESP	SCF	Mulliken	ESP	SCF	Mulliken	ESP
Imidazole	3.8 ± 0.4	3.118	3.508	2.129	3.575	3.861	2.412	3.869	3.535	2.213	3.494	4.025	2.855	3.962	3.855	2.822	3.810
Thiazole	1.61 ± 0.03	1.466	2.012	2.680	2.041	1.249	1.463	1.259	1.986	2.554	1.989	1.683	3.556	1.709	1.435	2.594	1.507
Furane	0.66	0.599	0.493	0.354	0.484	0.216	0.066	0.234	0.532	0.675	0.498	1.101	2.222	1.087	0.772	1.813	0.738
Methylsilane	0.735	–	0.374	0.276	0.331	0.432	0.175	0.402	–	–	–	0.702	1.572	0.238	0.672	0.027	0.658
NH_3	1.470	0.593	1.848	0.644	1.793	1.550	0.011	1.499	1.876	0.902	1.869	1.752	1.189	1.869	1.839	1.384	1.867
Dimethyl-ether	1.31	1.764	1.429	1.052	1.473	1.254	0.854	1.3194	1.847	1.181	1.384	1.847	3.109	1.901	1.475	2.512	1.531

the 3D electron density seems to be an excellent choice for generating a fairly correct picture of the electronic properties of any molecule. The mathematical technique underlying the ESP fit method involves least-squares fitting of the atomic charges to reproduce as closely as possible the charge density, which has been calculated quantum mechanically at a set of points in space surrounding the molecule. This yields much better results [9,11] than the Mulliken population analysis.

Whether a charge distribution obtained with a particular method is reliable and able to represent realistically the electronic proportions of a molecule must be checked against experimental data. One rather easy accessible experimental property is the molecular dipole moment. On the basis of atomic point charges a molecular dipole moment can be calculated in a simple and fast way and can be compared with appropriate experimental values which are listed for many compounds in literature (see for example [21]). Because the dipole moment depends crucially on the conformation of a molecule, only values for rigid molecules should be taken into consideration for comparative purposes. In order to decide on the applicability of a particular method for the calculation of charges in a series of molecules, one often proceeds by investigating not the entire flexible molecule but only small yet rigid fragments. Table 2.3 lists calculated and experimental dipole moments for a representative set of small and rigid structures. The dipole moments have been calculated using various methods and basis sets as well as the different procedures discussed earlier. The dipole moment is a quantum mechanically defined property; it can therefore also be calculated directly from wavefunctions (marked as SCF in Table 2.3). Corresponding results derived with a large basis set like 6–31G** are in especially good agreement with the experimental values.

Which type of procedure should be employed to investigate a particular molecular system depends on several factors. On the one hand the size of the molecules to be studied plays an important role; on the other the available computer power is the limiting factor for choosing a particular method.

Topological methods have the advantage over quantum chemical properties that they are very fast and give reasonable estimates of physical properties associated with charge. These methods generally produce dipole moments that are in good agreement with experimental values, partly a consequence of their calibration against experimental results. In contrast, the main disadvantage is the neglect of molecular geometries and conformations. Of course topological methods must fail in the case of molecules which contain atom types missing in the parameter list (see for example methylsilane in Table 2.3. Parameters for silicon are not included in the Gasteiger–Hückel method.).

Calculation of atomic charges from the molecular charge densities is the best choice if the results are for use in empirical energy functions for the purpose of interaction energy calculations. As can be deduced from Table 2.3 it is not absolutely necessary to use large ab initio basis sets. With smaller basis sets and even with the semi-empirical AM1 method dipole moments can be obtained which compare quite well with experimental values. However, the quality of the resulting dipole moment depends very distinctly on the procedure employed for generating the atomic point charges. All results obtained directly from molecular charge distribution are more realistic than the results of the Mulliken population analysis, which for some basis sets yields crude and erroneous dipole moments (see Table 2.3).

If a molecule of interest contains more than about 100 atoms then a sufficiently accurate calculation of the wavefunction is not feasible for the entire molecule. This impediment can be avoided by partitioning the large molecule into overlapping fragments. The fragment results then are transferred onto the large structure, hoping that the fragment properties correctly mirror the characteristics of the parent molecule.

However, even if point charges of high quality have been determined for a series of molecules these quantities are only weak arguments if the question of molecular similarity is the object of interest. Molecular similarity can be determined much more adequately on the basis of the 3D charge distribution. The most advantageous way to use this important and well-defined magnitude is through the MEPs.

2.4.1.2 Methods for Generating MEPs

MEPs are represented as interaction energies of a positively charged unit (a proton) with the charge density produced by the molecular set of nuclei and electrons at any point in space in the vicinity of the molecule. In general, a cut-off value is defined to limit the number of MEP points to be calculated. The MEP is a very useful tool in molecular modeling studies. It describes the electrostatic features of molecules and can be employed for the analysis and prediction of molecular interactions. For the generation of molecular electrostatic potentials two different approaches can be followed. The most desirable way is to calculate the MEPs directly from the quantum mechanically derived wavefunction. This procedure is straightforward and more accurate but time-consuming. A simpler approach is the calculation of MEPs on the basis of the atomic partial charges representing the molecular charge distribution. The MEP then is calculated applying the Coulomb equation for electrostatic interactions. Of course, the first procedure is by far superior and by all means should be used if sufficiently accurate wavefunctions are attainable for a particular molecule.

Many investigations are found in the literature which studied the basis set dependence of MEPs derived directly from wavefunctions [22–25]. It has also been shown that the electrostatic potential based on AM1 wavefunction correlates sufficiently well with ab initio results [22]. Therefore, AM1 can be used in all cases which cannot be handled due to molecular size at the ab initio level.

Visualization of MEPs

For the display of the molecular electrostatic potential different techniques are in use. The major obstacle for a fast and easy utilization of MEPs which permits the comparison of different molecules is the large amount of data points associated with this property. One very widely employed method to visualize MEPs is the display of the molecular electrostatic potential in the form of a 2D isocontour map in a particular plane of the molecule. The map may be displayed in color on a graphics screen, and can be manipulated in real-time. A single contour line represents values with similar energy. Regions containing a high nuclear contribution produce positive fields, corresponding to a repulsive interaction with a positive point charge, while those with a high electron density produce a negative potential, corresponding to an attractive interaction with a positive point charge.

The next level of complexity is reached by switching from 2D to 3D display mode. In principle, nothing changes since the molecule is completely wrapped by sets of isopotential shells. Each point on a particular shell experiences an electrostatic potential of the same sign and magnitude. With the help of this technique the over-all distribution of positively and negatively charged regions around a molecule can be visualized very distinctly. While 2D charts naturally may not always reveal a complete picture of the molecular electrostatic potential, the 3D isopotential surfaces effectively allow qualitative interpretation and comparison to be made between different compounds.

The third method for displaying the molecular electrostatic potential is associated with the calculation and visualization of the molecular surfaces. We will therefore dwell only shortly on the various definitions of the molecular surface. In the formal treatment of molecular surfaces the atomic positions are treated as points, whereas the electron clouds are approximated by spheres centered on the atomic centers. If the electron spheres are represented by the van der Waals radii, then the surface generated by summing all spheres is called the *van der Waals surface*. Van der Waals surfaces approximately represent the 3D volume requirements of molecules. A different type of surface which is often used in molecular modeling studies is the *solvent accessible surface*, also called *Connolly surface* [26]. The Connolly surface is the surface encircled by the center of a solvent probe as the probe molecule rolls over the van der Waals surface.

The electrostatic potential can be color-coded either onto the van der Waals or the Connolly surface. Each color at a defined surface point on the surface indicates a distinct energy value of the electrostatic potential. This technique attempts simultaneously to display both the shape of the molecules as well as their electrostatic properties. However, when larger molecules are studied the images become very complex. A solution to this problem is sometimes found by using the different techniques in a combined approach, since areas hidden in one display mode may be perceptible in the other (see Fig. 2.12).

The electrostatic potentials of different molecules, which bind to the same receptor site in a similar way, must share common features. It has been shown that in many cases where an atom-by-atom fit of the corresponding molecules does not lead to a satisfactory result, the MEP-directed superimposition yields an acceptable solution of the problem (see section 2.5.3).

As an example, it has been shown in a study of the electrostatic potential of histaminergic H_2 antagonists [27] that the imidazole ring of cimetidine and the guanidinothiazole ring of tiotidine can be superimposed on the basis of their electrostatic potential. This can be easily deduced from Fig. 2.13.

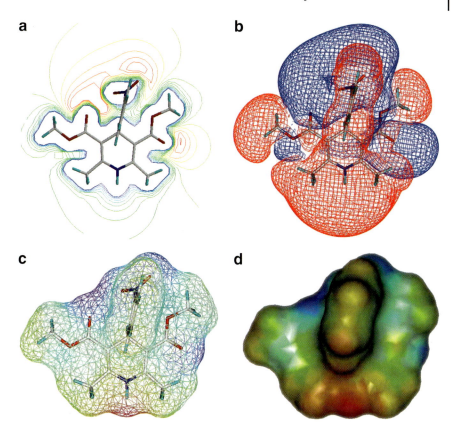

Fig. 2.12 Visualization of the molecular electrostatic potential (MEP) of nifedipine, using different techniques. a) The MEP is displayed as a 2-dimensional isocontour map in the plane of the dihydropyridine ring system. The electrostatic potential has been calculated directly from the ab initio wavefunction (using a 6–31G** basis set) and is contoured from −50 kcal mol^{-1} (red) to 90 kcal mol^{-1} (blue). b) The MEP is displayed in the form of isopotential surfaces. The electrostatic potential has been calculated by a point charge approach (ESP point charges have been derived from an ab initio calculation applying a 6–31G** basis set) and is displayed at the region of −5 kcal mol^{-1} (blue) as well as 5 kcal mol^{-1} (red). (The calculations have been performed using the quantum mechanical software package SPARTAN 3.0 [14]). c, d) The electrostatic potential displayed on the Connolly surface of nifedipine. The values of the electrostatic potential have been calculated using ESP derived point charges [the same as in (b)] and are displayed in the form of a simple dot surface as well as in the more sophisticated form of a solid "triangular" surface. Blue areas represent negative electrostatic potentials; red areas represent positive values. (The calculations have been performed using program MOLCAD [58]).

a

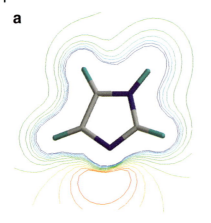

b

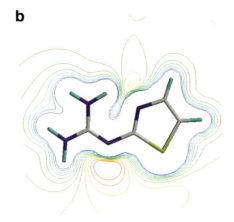

Fig. 2.13. Electrostatic potential of imidazole (a) and guanidinothiazole (b). The electrostatic potentials have been calculated using the ab initio wavefunction (with a 6–31G** basis set) and are contoured from −50 kcal mol^{-1} (red) to 90 kcal mol^{-1} (blue). (The calculations are performed using the quantum mechanical software package SPARTAN 3.0 [14]).The molecular electrostatic potential is a much more reliable indicator of electrostatic reactivity than the concept of atomic point charges. MEPs and their 3D representation have proven to be effective tools for analyzing and predicting the interaction of ligands with their macromolecular receptors.

2.4.2
Molecular Interaction Fields

Many biological processes are determined by non-covalent interactions between molecular structures. This is true for the docking of a ligand to a receptor, the interaction of a substrate with an enzyme, or the folding of a protein. Also in the world of crystals the non-covalent forces determine decisively the geometry and symmetry of the molecular arrangement. As a general rule binding only occurs if the generated energy of interaction overcomes the repulsive van der Waals forces. One method to investigate the energetic conditions between molecules approaching each other is the generation of molecular interaction fields. These fields describe the variation of interaction energy between a target molecule and a chemical probe moved in a 3D grid, which has been set around the target. The probes reflect the chemical characteristics of a binding partner, or fragments of it. By using computer graphics, molecular interaction fields can be displayed as 3D isoenergy contours. Contours of large positive energies indicate regions from which the probe would be repelled, while those of large negative energies correspond to energetically favorable binding regions.

The calculation of molecular interaction fields can be carried out using a variety of programs like GRID [28], CoMFA [29], HINT [30] or ISOSTAR/SUPERSTAR [31–33]. GRID is one of the most widely used programs for investigating molecular interaction fields. It works for small molecules as well as large protein molecules such as enzymes. Only Cartesian coordinates are needed as input. The list of probes is very comprehensive and the interaction energy is calculated on a regular grid of points surrounding the target molecule. The grid can also be confined to a particular fragment of the target molecule if only this part is of interest. The calculated energies are stored in a datafile and can be transferred for graphical display and analysis into most of the common molecular modeling programs [34–37]. 3D contour maps may then be generated at any selected energy level and studied together with the target molecule on a computer graphics system. The contouring is a quick process which allows the user to control the graphical results almost immediately.

In this chapter we will focus on the calculation of the interaction fields for small molecules; investigations of the fields for macromolecules will be discussed later (see section 4.6).

2.4.2.1 Calculation of GRID Fields

The probes which can be used for the calculation are small molecules, chemical fragments or particular atoms, e.g. a water molecule, a hydroxyl group or a calcium ion. These probes simulate the chemical characteristics of the corresponding binding partners, for example a potential receptor protein binding site or the neighbor molecule in a crystal. In the course of a GRID calculation the probe is moved systematically through a regular 3D array of grid points around the target structure. At each point the interaction energy between the probe and the target is calculated using the following empirical energy function:

$$E_{tot} = E_{vdw} + E_{el} + E_{hb} \tag{9}$$

where E_{tot} represents the total interaction energy, E_{vdw} represents the van der Waals interaction energy, E_{el} represents the electrostatic energy, and E_{hb} represents the interaction energy due to hydrogen bond formation.

The van der Waals interaction energy can be regarded as a combination of attractive and repulsive dispersion forces between non-bonded atoms. An atom of the probe is prevented from penetrating an atom in the target molecule by atomic repulsion and electron overlap. Repulsion forces can be estimated by an empirical energy function that becomes large and positive when the interatomic distance between two atoms is less than the sum of their van der Waals radii. The attractive part of the dispersion interaction is due to the correlated motion of electrons around the nuclei which results in induced dipole interactions. For non-polar molecules the balance between the attractive dispersion forces and the short-range repulsive forces can be described with the Lennard-Jones potential [37] (see for example Eq. (5) in section 2.2) which is implemented in the GRID program.

Electrostatic interactions are particularly important due to their long-range character for the attraction between ligand and macromolecular receptor.

The Coulomb equation (Eq. (6) in section 2.2) is widely used in molecular mechanics programs for the calculation of the electrostatic term because of its simple mathematical form. Its disadvantage is the fact that the heterogeneous media of molecular systems which consist of molecules with different dielectric properties are not sufficiently represented. The discontinuity between solute and solvent is taken into account by using an extended and more comprehensive form of Coulomb's law [28] which is used by GRID.

The directional properties of hydrogen bonds play a crucial role in determining the specifity of intermolecular interactions. It is therefore of utmost importance for a proper evaluation of interaction energies to describe this part of attractive forces between molecules in a correct form. A hydrogen bond can be regarded as intermediate-range interaction between a positively charged hydrogen atom and an electronegative acceptor atom [38]. The resulting distance between acceptor and donor atom is less than the sum of their van der Waals radii. In contrast to other noncovalent forces like dispersion and electrostatic point charge interactions, the hydrogen bonding interaction is directional, i.e. it depends on the propensity and orientation of the lone pairs of the acceptor heteroatom.

In order to comply with the requirements of these aspects the GRID method uses an explicit energy term for hydrogen bonds [39]. The functional form of this term has been developed to fit experimental data. All parameters are founded on experimental crystallographic data, i.e. direction, type and typical strength of these interactions are classified according to the real world of crystals.

The probes implemented in GRID are extensively defined by a variety of parameters, e.g. the hydrogen bonding possibility, the van der Waals radius or the atomic charge. The detailed description makes them very specific so that they can be regarded as realistic representatives of important functional groups found in the active site of macromolecules. As an example properties and parameters for three important probe groups are shown in Table 2.4.

Tab 2.4 Examples of the different parameters which are necessary to define GRID probe groups

	Methylprobe	Hydroxylprobe	Carboxylprobe
Van der Waals radius (Å)	1.950	1.650	1.600
Effective number of electrons	8	7	6
Polarizabilty ($Å^3$)	2.170	1.200	2.140
Electrostatic charge	0.000	−0.100	−0.450
Optimal hydrogen bond energy (kcal mol^{-1})	0.000	−3.500	−3.500
Hydrogen bonding radius (Å)	−	1.400	1.400
Number of hydrogen bonds donated	0	1	0
Number of hydrogen bonds accepted	0	2	2
Hydrogen bonding type	0	4	8

GRID also contains a table of parameters defining each type of atom that possibly exists in target molecules. The respective parameters define the strength of the van der Waals, the electrostatic and the hydrogen bond interactions potentially formed by an atom. The careful parametrization and the great variety of implemented probes make the GRID program a precise and widely used method for the investigation of interaction fields for small molecules as well as for macromolecular structures. The calculation of molecular interaction fields has been applied to a wide range of molecular modeling studies [40–45]. The strategy employed depends on the available structural information for ligands and macromolecular targets. If the 3D structure of a macromolecule is known, the interaction fields can be used to locate precisely favorable binding regions for the ligands. Subsequently these regions can be taken as a starting point for the design of new ligands for the particular receptor. This procedure will be described in detail in Chapter 3.

More often, situations are met where there is no structural information about the macro-molecular receptor and only the properties of the ligands are available. Under such circumstances molecular interaction fields can help to generate a more or less detailed picture of the potential receptor binding site. A prerequisite for this approach of course is that all investigated ligand molecules indeed bind to the same receptor site in an analogous mechanism. Only then can they be expected to exhibit a similar interaction pattern. Also, relative positions and size of the contours at any given energy level should be comparable. The energy level at which the contours must be compared is highly dependent on the probe type chosen.

Two different types of interaction fields for the Ca^{2+}-channel-blocking agent nifedipine are shown as an example in Fig. 2.14.

The interaction fields mark those parts of the corresponding binding region which possess particular chemical and physical properties. These properties can be translated into a model of the binding region of the macromolecule. If the macro-molecular target is a receptor protein the model is composed of single amino acid fragments which are located at the corresponding interaction regions. The amino acid fragments should satisfy the different binding regions which are common for all the active compounds. For example, the hydrophobic interaction fields possibly represent the location of hydrophobic amino acids such as phenylalanine, tryptophan, valine, leucine or isoleucine. Of course further investigations are necessary to specify the exact type of amino acid in each case.

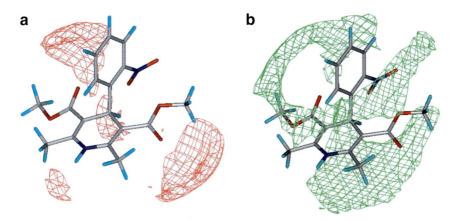

a   **b**

Fig. 2.14 Visualization of the molecular interaction fields of nifedipine. a) Favorable hydrogen-bonding regions derived from GRID [28] calculations using a hydroxyl probe (contour level: -3.5 kcal mol^{-1}). The favorable hydrophobic interaction regions obtained with a methyl probe are displayed in b) (contour level: -1.4 kcal mol^{-1}).

If a large set of compounds has to be studied it may become difficult to recognize all existing common interaction patterns. One way to solve this problem is to calculate the common interaction regions for different target molecules which were obtained in each case using the same probe. The common regions are mathematically detected in a gridpoint-by-gridpoint comparison of the fields. Only the hits are saved in a file and used for the generation of a common interaction field [46].

A more profound technique for comparison and analysis of molecular interaction fields is the use of chemometric methods [47–50]. These techniques will be discussed in detail in section 2.6.

2.4.2.2 Hydrophobic Interactions

As we have already discussed, attraction and repulsion between molecules are controlled by various types of interaction. One type which has not yet been considered is the so-called hydrophobic interaction. Hydrophobic interaction between molecules is a complex process that results primarily from entropic effects related to the change in the orientation of solvent molecules in the solvation shell wrapping the solute molecules, but also from the bulk form of the solvent molecules. For an effective hydrophobic interaction a close contact of the interacting hydrophobic surfaces is necessary [51,52]. The following piece of fiction might lead to a better understanding of hydrophobic bonding: a non-polar region of a deep binding cavity in a protein is not directly solvated. Nearby water molecules shield the cavity and are thought to form an iceberg-like structure which is stabilized by intermolecular hydrogen bonds between the water molecules. An interaction between the hydrophobic surfaces of the binding cavity and an entering substrate leads to a disruption of the ordered iceberg-like structure. The disruption yields an increase in entropy which results in a gain of free energy for the total system [52]. The desolvation of the substrate molecule also of course adds to the amount of newly formed bulk solvent molecules and must be taken into account. To date, the entropic effect usually is ignored in most modeling studies, because no simple method of calculation is available. On the other hand, it is generally accepted that the hydrophobic bonding or entropic effect does indeed play an important role in each drug–receptor interaction [53] as well as protein folding [54] event. As a natural follow-up the hydrophobic interaction should by all means be included in the energy balance of these processes.

Hydrophobicity can also be regarded as an empirical property of molecules encoding specific thermodynamic information about a molecule's interaction with its environment. Hitherto, several attempts were made for taking into account hydrophobic effects on the basis of experimental findings. The most important experimental measure of hydrophobicity is the solvent partition coefficient—expressed as log *P*—of a molecule between water and an organic phase. Since the log *P* can be determined experimentally it is a very useful tool. It can also be used to control and improve empirically developed methods, which are reported by several authors [55–57]. The prediction of log *P* can be achieved by transforming experimental solvent partition data for sets of variously substituted molecules into so-called hydrophobic fragment constants. These fragment constants represent the relative lipophilicity of particular structural elements found in the original set of molecules. The total lipophilicity of a compound (given by the log *P*) then can be calculated by summation of all fragment constants for a molecule under study. Today, fragment constants are available for a great variety of organic species with biological importance.

It should be noted that log *P* is a simple "one-dimensional" representation of hydrophobicity and only reflects an overall property. It is insufficient if a more detailed insight into molecular interactions between ligands and macromolecules is needed.

For that reason several attempts were made to utilize solvent partition coefficients as foundation to create a 3D representation for hydrophobicity. One way of approaching the problem is the generation of a hydrophobic field in analogy to the electrostatic field. This technique for example is implemented in the program HINT [30] or in the DRY probe in programm GRID [28].

The HINT model of hydrophobic interactions is based on the fact that solubility data can be regarded as just another physical property capable of mirroring the molecular interactions between solute and solvent molecules. In the framework of the HINT theory the fragment-level solvent partition data (the hydrophobic fragment constants) are reduced to hydrophobic atom constants [58]. These atomic descriptors are the key parameters and must be assigned for each atom in a molecule under investigation. Since the hydrophobic atom constants are derived from experimental partition experiments and solubility data, the obtained hydrophobic atom constants not only model the hydrophobic interactions but also include other types of molecular interactions, like electrostatic and van der Waals terms. The generated field therefore incorporates hydrophobic as well as hydrophilic parameters. It is called a *hydropathic field*. The calculation is performed using an empirical function (for a detailed description of the functional form, see [58]). The hydrophobic atom constants, the section of the solvent accessible surface created by each atom, and a distance function are included in the algorithm. The distance function is necessary to describe adequately the distance dependence of the hydrophobic effect in the biological environment. HINT generates 3D molecular grid maps in a similar way as discussed for comparable programs.

The result of a HINT study is a combined contour map for the hydrophobic and hydrophilic field around a molecule. Grid points with a positive sign represent a hydrophobic region. The opposite is true for hydrophilic (polar) segments of space. Because of the empirical nature of the data, it is difficult to decide at which energy niveau the fields have to be contoured. It is self-evident that the selected energy level directly determines the size of the visualized part of the field. For a proportionally correct balance in the size of the displayed contours it is usually advisable to contour the hydrophilic effect at a level 2–5 times higher than that of the hydrophobic effect [59].

The appearance of hydrophobic and hydrophilic fields again is demonstrated for the well-known Ca^{2+} channel blocker, nifedipine in Fig. 2.15(a).

The information obtained from the analysis of the hydropathic field can be exploited following different strategies. The qualitative information on the distribution of hydrophobic and polar properties in the vicinity of a series of molecules for example can be used to generate a 3D map of the unknown receptor macromolecule. If the investigated series is large and complex an interface allows the produced data set to be read directly into CoMFA for a more elaborate analysis [60].

If the structure of the macromolecular receptor is known, the generated hydropathic fields also can be used to optimize the structures of ligands for enhancement of the biological activity. For a review of other potential applications, see [60].

2.4.3
Display of Properties on a Molecular Surface

The display of hydrophobic and hydrophilic property distributions in the extramolecular space can also be projected onto a molecular surface. The program MOLCAD [61] employs for example the Connolly surface [26] of a molecule as a screen for mapping local molecular properties such as lipophilicity by a color-coded representation. A distance-dependent function must be defined in order to reflect correctly the influence of different atoms or fragments on the local lipophilicity at a certain point on the molecular surface. This can be realized for example by introducing a molecular lipophilicity potential [62], which can be regarded as a pendant to the molecular electrostatic potential. As in the case of the MEP, the projection of any local properties onto a surface facilitates the perception and interpretation of the distribution of the visualized property descriptor. The main advantage of the surface-bound representation of hydrophobicity is the fact that the analysis of large molecular systems, like proteins, is much easier in comparison with the evaluation of hydropathic fields. Because the theoretical background of both methods is equivalent, the results obtained should be comparable qualitatively. For both methods an effective test of reliability can be performed for all molecules for which experimentally derived log P values are available. However, the partition coefficient—like the charge distribu-

a

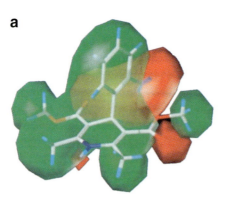

b

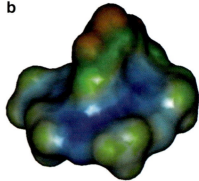

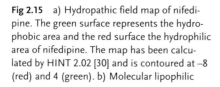

Fig 2.15   a) Hydropathic field map of nifedipine. The green surface represents the hydrophobic area and the red surface the hydrophilic area of nifedipine. The map has been calculated by HINT 2.02 [30] and is contoured at –8 (red) and 4 (green). b) Molecular lipophilic potential of nifedipine displayed on the Connolly surface. Brown areas on the surface represent more lipophilic parts and blue areas the hydrophilic parts of the molecule. (The calculation has been performed using program MOLCAD [61]).

tion—is drastically influenced by the conformation of a molecule. Moreover, the situation is further complicated by the conformation of a molecule being able to change when it migrates from the aqueous to the lipophilic environment. Unfortunately this fact limits the amount of test molecules to a rather small collection of rigid or at least semirigid structures. An example for the appearance of MOLCAD hydrophobic surfaces is shown in Fig. 2.15(b).

References

[1] Williams, D. E. Net Atomic Charge and Multipole Models for the ab Initio Molecular Electric Potential. In: *Reviews in Computational Chemistry*, Vol. 4. Lipkowitz, K. B., and Boyd, D. B. (Eds.). VCH: New York; 219–271 (1991).

[2] Gasteiger, J., and Marsili, M. *Tetrahedron* **36**, 3219–3228 (1980).

[3] Del Re, G. A. *J. Chem. Soc. London*, 4031–4040 (1958).

[4] Berthod, H., and Pullman, A. *J. Chem. Phys.* **62**, 942–946 (1965).

[5] Abraham, R. J., and Hudson, B. *J. Comput. Chem.* **6**, 173–185 (1985).

[6] Mullay, J. *J. Am. Chem. Soc.* **108**, 1770–1779 (1986).

[7] Mulliken, R. S. *J. Chem. Phys.* **23**, 1833–1840 (1955).

[8] Momany, F. A. *J. Chem. Phys.* **82**, 592 (1978).

[9] Cox, S. R., and Williams, D. E. *J. Comput. Chem.* **2**, 304–323 (1981).

[10] Singh, U. C., and Kollman, P. A. *J. Comput. Chem.* **5**, 129–145 (1984).

[11] Chirlian, L. E., and Francl, M. M. *J. Comput. Chem.* **8**, 894–905 (1987).

[12] Purcel, W. P., and Singer, J. A. *J. Chem. Eng. Data* **12**, 235–246 (1967).

[13] Frisch, M., Binkley, J. S., Schlegel, H. B., Raghavachari, K., Martin, R., Stewart, J. P., Bobrowicz, F., Defrees, D., Seeger, R., Whiteside, R., Fox, D., Fluder, E., and Pople, J. A. GAUSSIAN. Department of Chemistry, Carnegie Mellon University, Pittsburgh, Pennsylvania.

[14] SPARTAN, Schrödinger Inc., USA

[15] Schmidt, M. W., Boatz, J. A., Baldridge, K. K., Koseki, S., Gordon, M. S., Elbert, S. T., and Lam, B., GAMESS, Program No. 115, Quantum Chemistry Program Exchange, Indiana University, Bloomington, Indiana.

[16] Williams, D. E., and Yan, J. M. *Adv. Atomic Mol. Phys.* **23**, 87 (1988).

[17] Wiberg, K. E., and Rablen, P. R. *J. Comput. Chem.* **14**, 1504–1518 (1993).

[18] Reed, A. E., Weinstock, R. B., and Weinhold, F. A. *J. Chem. Phys.* **83**, 735–743 (1985).

[19] McWeeney, R. *Methods of Molecular Quantum Mechanics*, 2nd Ed. Academic Press: San Diego 1989.

[20] Destro, R., Bianchi, R., and Morosi, G. *J. Phys. Chem.* **93**, 4447–4457 (1989). Destro, R., Bianchi, R., Gatti, C., and Merati, F. *Chem. Phys. Letters* **186**, 47–52 (1991).

[21] McClellan, A. L. *Tables of Experimental Dipole Moments*, Vol. 2, Rahara Enterprise: El Cerrito, Californien, USA.

[22] Ferenczy, G. G., Reynolds, C. A., and Richards, W. G. *J. Comput. Chem.* **11**, 159–169 (1990).

[23] Rodriguez, J., Manaut, F., and Sanz, F. *J. Comput. Chem.* **14**, 922–927 (1993).

[24] Ford, G. P., and Wang, B. *J. Comput. Chem.* **14**, 1101–1111 (1993).

[25] Price, S. L., Harrison, R. J., and Guest, M. F. *J. Comput. Chem.* **10**, 552–567 (1989).

[26] Connolly, M. L. *Science* **221**, 709–713 (1983).

[27] Höltje, H.-D., and Batzenschlager, A. *J. Comput.-Aided Mol. Design* **4**, 391–402 (1990).

[28] Goodford, P. J. *J. Med. Chem.* **28**, 849–857 (1985).

[29] Cramer, R. C., Patterson, D. E., and Bunce, J. D. *J. Am. Chem. Soc.* **110**, 5959–5967 (1988).

[30] Kellogg, G. E., Semus, S. F., and Abraham, D. J. *J. Comput.-Aided Mol. Design* **5**, 545–552 (1991).

[31] Bruno, I. J., Cole, J. C., Lommerse, J. P., Rowland, R. S., Taylor, R., Verdonk, M. L. *J. Comput.-Aided Mol. Design* **11**, 525–537 (1997).

[32] Verdonk, M. L., Cole, J. C., Taylor, R. *J. Mol. Biol.* **289**, 1093–1108 (1999).

[33] Boer, D. R., Kroon, J., Cole, J. C., Smith, B., Verdonk, M. L. *J. Mol. Biol.* **312**, 275–287 (2001).

[34] INSIGHT/DISCOVER, Accelrys Inc., San Diego, California, USA.

[35] SYBYL, Tripos Associates, St. Louis, Missouri, USA.

[36] MACROMODEL, Mohamadi, F., Richards, N. G. C., Guida, W. C., Liskamp, R., Carfield, C., Chang, G., Hendrickson, T., and Still, W. C. *J. Comput. Chem* **11**, 440–464 (1990).

[37] Lennard-Jones, J. E. *Proc. Roy. Soc.* **106A**, 463–477 (1924).

[38] Dean, P. M. *Molecular Foundations of Drug–Receptor Interaction.* Cambridge University Press: Cambrige 1986.

[39] Wade, R. C. Molecular interaction fields. In: *3D QSAR in Drug Design – Theory, Methods and Applications*, Kubinyi, H. (Ed.). ESCOM Science Publishers B. V.: Leiden; 486–505 (1993).

[40] Wade, R. C., Clark, K. J., and Goodford, P. J. *J. Med. Chem.* **36**, 140–147 (1993).

[41] Reynolds, C. A., Wade, R. C., and Goodford, P. J. *J. Mol. Graphics* **7**, 103–108 (1989).

[42] Meng, E. C., Shoichet, B. K., and Kuntz, I. D. *J. Comput. Chem.* **13**, 505–524 (1992).

[43] Höltje, H.-D., and Jendretzki, U. *Pharm. Pharmacol. Lett.* **1**, 89–92 (1992).

[44] Wade, R. C., and Goodford, P. J. *Br. J. Pharmacol. Proc. Suppl.* **95**, 588 (1988).

[45] Cruciani, G., and Watson, K. A. *J. Med. Chem.* **37**, 2589–2601 (1994).

[46] Höltje, H.-D., and Anzali, S. *Die Pharmazie* **47**, 691–697 (1992).

[47] Baroni, M., Costantino, G., Cruciani, G., Riganelli, D., Valigi, R., and Clementi, S. *Quant. Struct.-Act. Relat.* **12**, 9–20 (1993).

[48] Wold, S., Johansson, E., Cocchi, M. PLS – Partial Least-Squares Projections to Latent Structures. In: *3D QSAR in Drug Design – Theory. Methods and Applications.* Kubinyi, H. (Ed.). ESCOM Science Publishers B. V.: Leiden; 523–550 (1993).

[49] Klebe, G., and Abraham, U. *J. Med. Chem.* **36**, 70–80 (1993).

[50] Folkers, G., Merz, A., and Rognan, D. CoMFA: Scope and Limitations. In: *3D-QSAR in Drug Design – Theory, Methods and Applications.* Kubinyi, H. (Ed.). ESCOM Science Publishers B.V.: Leiden; 583–618 (1993).

[51] Tandford, C. *Science* **200**, 1012–1018 (1978).

[52] Tandford, C. *The Hydrophobic Effect.* Wiley: New York 1980.

[53] Suzuki,T., and Kudo, Y. *J. Comput.-Aided Mol. Design* **4**, 155–198 (1990).

[54] Nicholls, A., Sharp, K. A., and Honig, B. *Proteins* **11**, 281–296 (1991).

[55] Hansch, C., and Fujita, T. *J. Am. Chem. Soc.* **86**, 1616–1626 (1964).

[56] Rekker, R. F., and Mannhold, R. *Calculation of Drug Lipophilicity.* VCH: Weinheim 1992.

[57] Ghose, A. K., and Crippen, G. M. *J. Comput. Chem.* **7**, 565–577 (1986).

[58] Kellogg, G. E., Joshi G. S., and Abraham, D. J. *Med. Chem. Res.* **1**, 444–453 (1992).

[59] Kellogg, G. E., and Abraham, D. J. *J. Mol. Graphics* **10**, 212–217 (1992).

[60] Abraham, D. J., and Kellogg, G. E. Hydrophobic Fields. In: *3D QSAR in Drug Design – Theory, Methods and Applications.* Kubinyi, H. (Ed.). ESCOM Science Publishers B. V.: Leiden; 506–522 (1993).

[61] Heiden, W., Moeckel, G., and Brickmann, J. *J. Comput.-Aided Mol. Design* **7**, 503–514 (1993).

[62] Furet, P., Sele, A., and Cohen, N. C. *J. Mol. Graphics* **6**, 182–189 (1988).

2.5
Pharmacophore Identification

2.5.1
Molecules to be Matched

In the first sections of this book we have described how physico-chemical character-istics of molecules can be calculated and visualized. Now, we will discuss how this knowledge can be used to understand or predict the pharmacological properties of a compound. In the large majority of cases the basis for a pharmacodynamic effect is the interaction of a certain substance with a protein of physiological importance. The macromolecule might be an enzyme or a receptor. In both cases there must exist a highly specific 3D cavity which serves as binding site for the drug molecule. Compounds exerting qualitatively similar activities at the same enzyme or receptor therefore must possess closely related binding properties. That is, these molecules must present to the macromolecular binding partner structural elements of identi-cal chemical functionality in sterically consistent locations. In short, congeners of a defined pharmacological group possess an identical pharmacophore, and one of the major tasks to be solved using molecular modeling techniques is the determination of pharmacophores for congeneric groups of drug molecules. Because the 3D struc-tures of most receptors hitherto remain undiscovered, information on the corre-sponding hypothetical pharmacophore as a matter of fact is a very important source for understanding drug–receptor interactions at the molecular level.

When all physico-chemical properties have been intensively studied the question remaining is "How do we have to superimpose the members of a series to find the pharmacophore?" In order to answer this question we have first to define the phar-macophoric elements. That is, we must decide what functional groups or atoms have to be superimposed. Of course this question cannot be answered completely objectively in an automatic procedure because one always has to decide in advance on the atom pairs which correspond between two molecules. This may produce a large number of useless data if known structure–activity relationship information is not included. This facilitates the superpositioning procedure, because it drastically limits the number of solutions. It should be noted that similarity between ligands must not comprise the whole molecule, because most of the ligand molecules are not completely wrapped by receptor binding sites when they are bound to it. This also reduces the number of reasonable solutions.

If hydrogen bonds are supposed to be important for the pharmacophore then the direction and distance of lone pairs should be added to the atomic pattern of the molecules under study. This can be realized for example by locating dummy atoms in corresponding positions. These positions then are labelled by different flags as hydrogen bond-acceptor or -donor sites (only hydrogens bound to heteroatoms) and can be used as a first test for a superposition mode (for example, the program AUTOFIT [1]). Furthermore, planar elements like aromatic ring systems can be trea-ted as special structural units. In this case for example the center of the ring system can be defined as matching point instead of the ring system. Other planar groups can be handled analogously.

If the set of molecules contains only very flexible congeners then the search for a common pharmacophore is not only very difficult and tedious but also might even yield either none or an arbitrary (and therefore useless) result. This task can be easily performed and is of far greater significance if rigid or at least semi-rigid compounds are present. These of course must be highly active, otherwise they can not be used as matrices for the flexible ligands. By the same token, the consideration of highly active but conformationally restricted molecules relieves the need to prove that one is indeed dealing with bioactive conformations.

The selection of the molecules to be superimposed is very important if significant results are to be obtained. The easiest to perform, but rather ineffective, case is the superimposition of structurally very similar compounds. This does not provide much information, so it is much more effective to include in the series structures containing different skeletons. As a natural follow-up this leads to a situation which is highly desirable where a simple atom-to-atom superpositioning is not possible but a matching of functionally equivalent elements or a matching of molecular fields must be performed.

One further point must be addressed. Are inactive molecules or molecules with only low activity to be taken into consideration? It seems useful at first to superpose highly active molecules alone. The derived pharmacophore then can be tested against, and eventually modified by, inclusion of low active and inactive congeners. The same holds true for antagonists and agonists of one receptor type. Superpositions should be performed for both groups separately. However, both models subsequently can possibly be combined, because very often competitive antagonists are bound at least partially in the agonistic receptor binding site. However, it should be noted that overlapping binding sites of agonists and antagonists are indeed common but do not necessarily exist.

One also has to keep in mind that indirect approaches suffer from serious limitations. First, all ligands must bind to the target protein at the same location and preferably adopt the same binding mode. If the former prerequisite is missing, the superposition will be misleading (for example see [2, 3]). Further, pharmacophore models are usually restricted to low-energy conformations. Since the number of accessible conformations of a molecule increases dramatically with the allowed energy range, it is impossible to consider all conformations for pharmacophore generation. Hence, strained conformations, such as observed in the transition state of a chemical reaction, will usually not be covered by pharmacophore models. A detailed comparison between the energies of in silico generated and experimentally determined protein-bound ligand conformations has been published recently [4, 5].

Several different superpositioning procedures are available. They comprise manual or automatic fitting by rigid-body rotation or flexible-fitting procedures where both root mean square (rms) derivation between the fitted atom pairs and conformational energies are minimized. Other important superpositioning techniques perform alignments on the basis of equivalences detected in molecular surfaces or molecular field properties.

2.5.2
Atom-by-Atom Superposition

The least-squares technique for superpositioning of corresponding atom positions is the most widely used method. Two molecules are superimposed by minimizing the rms of the distances between the corresponding atom pairs in the molecules. The rms value represents a measure for the quality of the fit. This procedure is very powerful in discovering dissimilarity between molecules which seem to be apparently similar. The weak point is that it is required to decide in advance which atom pairs match. It is obvious that different superpositions are obtained depending on the atoms used for the procedure. The method cannot be applied to molecular systems in which atom-to-atom correspondences are not detectable in advance. However, rigorous similarity in atomical structure is not a prerequisite for the interaction of different molecules with the same receptor. Therefore for a large number of cases where pharmacological data and structure–activity studies urge upon a common mechanism of action for a set of dissimilar molecules the conventional least-squares superpositioning method is considered inadequate.

One may try to escape such a situation by performing a manual, interactive superposition if the test set is not too large. In principle, any number of molecules can be investigated directly on the graphics display and the fit may be judged visually. Certainly this procedure is very creative and may stimulate new ideas about the underlying mechnism of experimentally detected structure–activity data. On the other hand, such a procedure naturally is biased and often cannot correctly be reproduced, because a computational optimization is not applicable.

An efficient and fast search technique which can be used very successfully for the generation of pharmacophore models, the Active Analogue Approach, was developed by Marshall et al. [6,7]. This technique utilizes a systematic search algorithm for calculating a representative number of sterically and energetically allowed conformations for congeneric molecules. For each of these conformations a set of distances between pharmacophoric groups believed to be important for the interaction with the receptor is generated. If each set of distances for one molecule is compared with all sets of all the other molecules—with the intention to find possible correspondences—the problem would not be solvable except for small molecules. On the other hand, in the framework of pharmacophore identification the complete conformational space of all compounds is not of interest but rather only those subregions which are accessible to all active ligands. As we have discussed earlier, it is of major advantage to include rigid or semi-rigid compounds in a conformational analysis for a series of flexible molecules. For that reason the conformational search is started with the most rigid molecule. After determination of the respective distance map for this compound these distances are used as constraints in the conformational search runs for the more flexible molecules. Following these lines the results of a search on one active and rigid analog are taken as a basis to explore the conformational space of all the other congeners of the series. As all of the active compounds must fit the receptor model the search is restricted to those regions of conformational space which correspond to the previously defined model. For example, according to the model if a pair of atoms must lie within a certain distance range in order to agree with the constraints, then only those torsions that will allow this constraint to be satisfied need to be calculated. An example which has demonstrated the strength of the Active Analogue Approach dealt with 28 angiotensin-converting enzyme (ACE) inhibitors in an effort to predict a model for the ACE active site [8]. Applying this technique the search time was reduced by more than three orders of magnitude in comparison with a previously performed conventional systematic search study on the same subject.

Another mapping procedure, which in contrast does not use an explicit atom-by-atom superposition approach is SEAL [9]. This program allows a rapid pairwise comparison of dissimilar molecules. The similarity score, as an indicator of the quality of fit, is calculated from a summation over all possible atom pairs between the two molecules. Each atom pair is weighted by the relative distance between the contributing atoms. In doing so the alignment function considers all theoretically possible atom pairs in the molecules in the comparison procedure and not only one atom pair, as in the atom-by-atom fit approach. As a consequence the resulting superposition reflects to some extent the properties associated with the global shape of the molecules. The program also offers the possibility to include physico-chemical properties in the alignment procedure. Therefore, the terms used in the pairwise summation can be composed from any physico-chemical quality supposed to be important for the biological effect. In the original version the authors used only van der Waals radii as an expression of sterical volume as well as point charges mimicking the electronic molecular properties to optimize the alignment.

An extended version of the original SEAL program has been developed by Klebe et al. [10]. Several structural alignment methods including rigid-body superposition based on an efficient overlap optimization are provided. Different molecular fields are described by sets of Gaussian functions. Consideration of the intramolecular conformational strain energy, as well as a flexible-fitting routine is also available [11].

There also exist mapping techniques which include as one of the first steps in the computational protocol the automatic and therefore unbiased identification of atomic centers or site points as correspondences used for superpositioning. Site points may include points of the molecular surface representing molecular features like hydrogen bond acceptor or donor characteristics. Several commercial program packages like CATALYST [12] offer such functionalities. Others like DISCO [13], RECEPS [14], and AUTOFIT [1] have been thoroughly discussed in a recent review of common alignment programs [15]. As described earlier the superposition is performed by matching the assigned corresponding atoms or site points in all possible combinations.

Most of the modern alignment programs treat the molecules to be superimposed as flexible, whereas one rigid reference compound is required. One popular program, FlexS [16], uses a flexible superpositioning on the basis of a combinatorial matching procedure. Pairs of molecules are aligned, one of which is considered rigid, allowing the other molecule to be flexibly fitted. The FlexS method originated from the related docking program FlexX [17]. FlexS tries to decompose the flexible structure into relatively rigid portions, to begin with the alignment of the corresponding pharmacophoric features [18]. The remaining portions of the molecule are incorporated in an iterative incremental procedure. The similarity between the superimposed molecules is calculated on the basis of energy-like matching terms for paired intermolecular interactions and overlap terms of the utilized Gaussian functions. The approach has been validated on experimental data obtained from X-ray structures [18]. Due to its computational efficiency the method is extremely fast and can be used to screen large databases of compounds.

2.5.3
Superposition of Molecular Fields

Since molecules recognize each other by characteristic properties on or outside their van der Waals volume—and not through their atomic skeleton—the determination of molecular similarity should take into account the molecular fields. As a natural follow-up the superpositioning approaches should also concentrate on mapping and comparing these properties. For matching purposes the molecules are located in a 3D grid of equally spaced field points. Each grid point is loaded with a certain characteristic property measure such as charge distribution, hydrophobic potential or simply information on the size of the volume. Similarity thresholds can be defined in order to guide the optimization procedure to a significant and unequivocal result. Single grid points or clusters of adherent grid points can be assigned different weights in order to pay as close as possible attention to structure–activity relationships. One molecule—preferentially with limited conformational freedom—is chosen as the template molecule. The grid loadings of the template serve as a measure for the various properties and all trial molecule grids are manipulated by rotation and translation to find the best fit of the grid values. The computational technique of orientational search which has to be used is extremely time-consuming. Different methods have been described which mirror different levels of complexity but also utilize various field properties. Manaut et al. [19] reported an effective method which maximizes the similarity between molecular surfaces on the basis of the molecular electrostatic field. Other groups such as Clark et al. [20] or Dean et al. [21] use physico-chemical field properties calculated using Lennard-Jones potentials, or replace the regular grid-based evaluation technique by an integration over Gaussian-type functions to approximate the electrostatic potential. Goodness-of-fit indices can be calculated for example as ratio of the number of commonly occupied grid points to the total number of grid points.

In summary, the tools for matching molecular surfaces do exist. Since the corresponding methods do not require any atom correspondences between molecules, they can be used efficiently for superposing dissimilar molecules. However, this might become a routine technique only if the complicated calculations can be made fast enough to deal with a large number of conformations for each molecule to be superimposed. For a detailed comparison between atom-by-atom and field-based methods the reader is referred to the literature [15,22,23]. An outstanding collection of articles and reviews on the subject can be found in [24].

References

[1] Kato, Y., Inoue, A., Yamada, M., Tomioka, N., and Itai, A. *J. Comput.-Aided Mol. Design* **6**, 475–486 (1992).

[2] Klebe, G., Abraham, U. *J. Med. Chem.* **36**, 70–80 (1993).

[3] Böhm, H. J., Klebe, G. and Kubinyi, H. *Wirkstoffdesign*, Spektrum, Akad. Verl. (1996).

[4] Boström, J. *J. Comput.-Aided Mol. Design* **15**, 1137–1152 (2001).

[5] Boström, J., Norrby, P.-O. and Liljefors, T. *J. Comput.-Aided Mol. Design* **12**, 383–396 (1998).

[6] Marshall, G. R., Barry, C. D., Bosshard, H. E., Dammkoehler, R. A., and Dunn, D. A. The conformational parameter in drug design: The active analog approach. In: *Computer-Assisted Drug Design*, ACS Monograph 112. Olsen, E. C., and Christoffersen, R. E. (Eds.). American Chemical Society: Washington D. C. 205–226 (1979).

[7] Dammkoehler, R. A., Karasek, S. F., Shands, E. F. B., and Marshall, G. R. *J. Comput.-Aided Mol. Design* **3**, 3–21 (1989).

[8] Mayer, D., Naylor, C. B., Motoc, I., and Marshall, G. R. *J. Comput.-Aided Mol. Design* **1**, 3–16 (1987).

[9] Kearsley, S. K., and Smith, G. M. *Tetrahedron Comput. Methodol.* **3**, 615–633 (1990).

[10] Klebe, G., Mietzner, T., Weber, F. *J. Comput.-Aided Mol. Design* **8**, 751–778 (1994).

[11] Klebe, G., Mietzner, T., Weber, F. *J. Comput.-Aided Mol. Design* **13**, 35–49 (1999).

[12] CATALYST Accelrys Inc., San Diego, USA.

[13] Martin, Y. C., Bures, M. G., Danaher, E. A., DeLazzer, J., Lico, I., and Pavlik, P. A. *J. Comput.-Aided Mol. Design* **7**, 83–102 (1993).

[14] Kato, Y., Itai, A., and Iitaka, Y. *Tetrahedron* **43**, 5229–5236 (1987).

[15] Lemmen, C. and Lengauer, T. *J. Comput.-Aided Mol. Design* **14**, 215–232 (2000).

[16] Lemmen, C. and Lengauer, T. *J. Comput.-Aided Mol. Design* **11**, 357–368 (1997).

[17] Rarey, M., Kramer, B., Lengauer, T. and Klebe, G. *J. Mol. Biol.* **261**, 470–489 (1996).

[18] Lemmen, C., Lengauer, T. and Klebe, G. *J. Med. Chem.* **41**, 4502–4520 (1998).

[19] Manaut, M., Sanz, F., Jose, J., and Milesi, M. *J. Comput.-Aided Mol. Design* **5**, 371–380 (1991).

[20] Clark, M., Cramer III, R. D., Jones, D. M., Patterson, D. E., and Simeroth, P. E. *Tetrahedron Comput. Methodol.* **3**, 47–59 (1990).

[21] Dean, P. M. Molecular recognition: The measurement and search for molecular similarity in ligand–receptor interaction. In: *Concepts and Applications of Molecular Similarity*. Johnson, M. A., and Maggiora, G. M. (Eds.). Wiley: New York; 211–238 (1990).

[22] Mason, J. S., Good, A. C. and Martin, E. *J. Curr. Pharm. Design* **7**, 567–597 (2001).

[23] Good, A. C., Mason, J. S. *Three-Dimensional Structure Database Searches*. In: Reviews in Computational Chemistry, Vol. 7. Lipkowitz, K. B., and Boyd, D. B. (Eds.). VCH: New York; 73–95 (1995).

[24] Kubinyi, H. (Ed.) *3D QSAR in Drug Design. Theory and Applications.* ESCOM: Leiden, (1993).

2.6
3D QSAR Methods

3D QSAR (three-dimensional quantitative structure-activity relationship) techniques are the most powerful computational means to support the chemistry side of indirect drug design projects. The primary aim of these techniques is to establish a correlation of the biological activities of a series of structurally and biologically characterized compounds with the spatial fingerprints of numerous field properties of each molecule, such as steric demand, lipophilicity and electrostatic interactions. Typically, a 3D QSAR study allow to identify the pharmacophoric arrangement of molecular fragments in space and provides guidelines for the design of the next generation of compounds with enhanced biological potencies.

The number of 3D QSAR studies has exponentially increased over the last decade, since a variety of methods are commercially available in user-friendly, graphically guided software [1–3]. Besides the commercial distribution, a major factor in the continuing enthusiasm for 3D QSAR comes from the proven ability of several of these methods to predict correctly the biological activity of novel compounds [4]. However, the ease of application of 3D QSAR programs may encourage especially novice modelers to apply the methods to all available data sets. It is the aim of this chapter to give an introduction to relevant 3D QSAR methods, but a special interest will be also taken in the limitations of the various approaches.

2.6.1
The CoMFA Method

The CoMFA [1] method (Comparative molecular field analysis) was developed as a tool to investigate three-dimensional quantitative structure-activity relationships. 3D QSAR approaches use statistical methods (chemometrical methods) to correlate the variation in biological or chemical activity with information on the three-dimensional structure for a series of compounds. A CoMFA analysis starts with traditional pharmacophore modeling to suggest a bioactive conformation of each molecule and ways to superimpose the molecules under study. This is not a trivial task, as has been documented in the previous section. The underlying idea of the comparative molecular field analysis is that differences in a target property, e.g. biological activity, are often closely related to equivalent changes in the shapes and strengths of the noncovalent interaction fields surrounding the molecules. To state it differently, the steric and electrostatic fields provide all the necessary information to understand the biological properties for a set of compounds. As in the GRID approach, the molecules are located in a cubic grid and the interaction energies between the molecule and a defined probe are calculated for each grid point. Usually, only two potentials, the steric potential, expressed in a Lennard-Jones function, and the electrostatic potential, expressed in a simple Coulomb function, are used within a CoMFA study. It is obvious that the description of molecular similarity is not a trivial task, nor is the description of the interaction process of ligands with their biological targets. In the standard application of CoMFA, the potentials chosen provide only enthalpic contributions of the free energy of binding [5]. However, many binding effects are governed by hydrophobic and entropic contributions. Therefore, one has to characterize in advance the main contributions of forces expected and to judge whether CoMFA will actually be able to find realistic results under these conditions.

2.6.1.1 Biological Data Used for 3D QSAR Studies

As with any QSAR method, an important point is the question if the biological activities of all compounds studied are of comparable quality. Preferably the biological data have been obtained in the same laboratory under identical conditions. All compounds being tested in a system must have the same mechanism (binding mode) and all inactive compounds must be shown to be truly inactive. Only in vitro data should be considered, since only in vitro experiments are able to reach a real equilibrium. Other test systems undergo time-dependent changes by multiple coupling to parallel biochemical processes, for example membrane permeation. Another critical point is the existence of transport phenomena and diffusion gradients influencing all biological data. One has to bear in mind that CoMFA was developed to describe only one interaction step in the lifetime of ligands. In all cases where nonlinear phenomena result from drug transport and distribution, any 3D QSAR technique should be applied with caution.

Ideally, the biological activities of the molecules used for a CoMFA study should span a range of at least three orders of magnitude. For all molecules under study, the exact 3D structure must be known and reported. If no information about the exact stereochemistry of the tested compounds is available, as is the case for mixtures of enantiomers or diastereomers, it is not possible to include these compounds in a CoMFA study.

2.6.1.2 Deriving the CoMFA Model

Once the biological activities have been tested and the molecules have been superimposed in their putative bioactive conformation, the CoMFA analysis continues by calculating the intermolecular interaction fields surrounding each molecule. For CoMFA this implies that a lattice surrounding all the molecules is constructed, and the electrostatic and van-der-Waals energies with a chosen probe atom are calculated at each intersection of the lattice. As a default, extending the lattice by 4 Å beyond any dimension of any molecule in the dataset is adequate for most of the CoMFA analyses [4]. The usual CoMFA calculation is performed using a lattice spacing of 2 Å. A controversial discussion about lattice spacing can be found in the literature [6]. Superior results are often obtained when using a 2 Å spacing as opposed to the more accurate 1 Å spacing. In addition, the CoMFA program provides a variety of other parameters (probe atoms, charges, energy scaling, energy cutoffs, etc.) which can be adjusted by the user. This flexibility in parameter settings enables the user to fit the whole procedure as closely as possible to the particular problem. However, this flexibility enhances the possibility of chance correlations. Interestingly, nearly all of the successful CoMFA analyses have been performed with the default parameters. It is beyond the scope of this chapter to discuss all the consequences of modifying CoMFA parameters. For this purpose the reader is referred to two articles dealing with an extensive analysis of parameter settings and their influence on the CoMFA model [6, 7].

2.6.1.3 Statistical Quality of CoMFA Models

The relationships between the biological activities and the generated interaction fields are evaluated by the multivariate statistical technique of partial least squares (PLS) calculation. For a detailed description of the underlying mathematics of multivariate analysis the reader is referred to the literature [8, 9]. The PLS method is able to build a statistical model even though there are more energy values than compounds, because the various energy values are correlated with each other and many are unrelated to biological activity. By these assumptions, the PLS method is able to extract a weak signal dispersed over many variables. Generally, no more than five or six linear combinations of the energy values are needed to build a realistic model. Since PLS operates on many variables (interaction energies), a realistic concern is that the data will be over-fitted. For this reason PLS models are validated by so-called „leave-one-out" cross-validation. This procedure involves calculating as many models as there are data points (molecules) in the data set. For each model in turn, one of the compounds is left out and its activity is predicted from the model lacking the compound. After each compound has been predicted once, the Q^2 value (square of the cross-validated correlation coefficient, see Eq. 10) and the SDEP value (standard deviation of the error of prediction, see Eq. 11) are calculated from the observed and predicted potencies of each compound.

$$SDEP = \sqrt{\sum \frac{(Y-Y')^2}{N}} \tag{10}$$

$$Q^2 = 1 - \frac{\sum(Y-Y')^2}{\sum(Y-\bar{Y}')^2} \tag{11}$$

Y: experimental value, Y':predicted value, $\bar{Y}'$: average value, N: number of objects

The SDEP value generally decreases for the first few latent variables, reaches a minimum, and then increases to indicate over-fitting of the data. The number of latent variables considered should be determined carefully. If adding a variable decreases the SDEP by less than 5%, the simpler model is preferred, because it contains most of the signal in fewer variables. Considering more variables results in adding more noise to the model. In CoMFA a Q^2 value above 0.3 is usually considered statistically significant and acceptable [6]. However, several studies indicated that the statistical significance of CoMFA models should be carefully examined. To investigate the risk of chance correlation, a scrambling-test is usually performed. In this test, the biological activities are randomly distributed among the molecules in the training set. On the basis of these randomly assigned activities novel CoMFA models are generated and the Q^2 values are compared with the original model. To investigate the risk of chance correlation, Krystek et al. used scrambled biological activities to test their CoMFA model for 36 endothelin subtype-A (ET$_A$) receptor ligands [10]. Scrambled biological activities yielded a model, using a single latent variable, with a Q^2 of 0.43 compared to a Q^2 value of 0.70 for the model with correctly assigned activities. The R^2 values of randomly and correctly generated models were comparable, indicating that the R^2 value of a CoMFA model cannot be considered for validation. These results also demonstrated that it is not possible to provide an exact Q^2-cutoff value for a chance correlation. From our experience, robust and predictive models should have Q^2 values above 0.5 [7, 11–13].

To overcome the problem of chance correlation, several other strategies have been suggested [6]. For example, more robust cross-validation techniques may be applied. Using these methods, 10% or 20% of the compounds are randomly selected and a model is generated with the remaining 90% or 80% of the compounds, which are then used to predict the randomly selected compounds. To obtain statistically reliable results, this procedure is repeated several times. The „leave-groups-out" cross-validation has been shown to yield better indices for the robustness of a model than the usual „leave-one-out" procedure [14].

2.6.1.4 Interpreting the Results

Another way to analyze the results of a CoMFA study is the visual inspection of the model. For this purpose, the contours which represent 3D locations of fields with significant contribution to the model are displayed. Steric and electrostatic contributions are contoured separately and shown in different colours. The steric contours are relatively easy to interpret. Positive contours show regions in space that, if occupied, increase potency, negative contours those that decrease potency. Interpretation of the electrostatic maps is more complicated because of the electroneutrality requirement and because either positive or negative charges in electrostatics can increase potency. If a CoMFA analysis shows significant electrostatic effects, the user must examine the underlying electronic effects of corresponding functional groups to establish which is the true effect and which is an artificial correlation.

Normally, CoMFA contour maps are not considered to mirror exactly the corresponding attributes of the target protein, and such a comparison should only be drawn with extreme care. However, when the ligand alignment is based on the geometry of protein-bound conformations, the CoMFA steric and electrostatic coefficient contours may correspond to some extent to the steric and electrostatic environment of the binding site. For example, Oprea et al. [16] used inhibitor-bound enzyme X-ray structures not only to align the molecules, but also to evaluate the CoMFA results by comparing the CoMFA coefficient contour maps with the binding site structure. Several residues that are important to ligand binding were found to have corresponding steric and electrostatic CoMFA fields. However, the comparison also revealed some limitations of the model, since not all key residues displayed overlap with the CoMFA fields. Similar observations have been made in our CoMFA studies [11, 12].

2.6.2
CoMFA-related Methods

2.6.2.1 CoMSIA

Due to the problems associated with the functional form of the Lennard-Jones potential used in most CoMFA methods [15], Klebe et al. [2] have developed a similarity indices-based CoMFA method, which is named CoMSIA (Comparative Molecular Similarity Indices Analysis). The method uses Gaussian-type functions instead of the traditional CoMFA potentials. Three different indices related to steric, electrostatic and hydrophobic potentials were used in their study of the classical Tripos steroid benchmark dataset. Models of comparable statistical quality with respect to internal cross-validation of the training set, as well as predictivities of the test set, were derived using the CoMSIA method. The clear advantage of this method lies in the functions used to describe the molecules studied, as well as in the resulting contour maps. The contour maps obtained with CoMSIA are easier to interpret compared to the maps obtained with the CoMFA approach. The CoMSIA procedure also avoids the cutoff values used in CoMFA to prevent the potential functions from assuming unacceptably large values. For a detailed description of the method as well as its application the reader is referred to the literature [17, 18]. Recently, the authors of CoMSIA have included a novel hydrogen-bond descriptor which should overcome the problem of underestimating hydrogen bonds in CoMFA studies [19].

2.6.2.2 GRID and GOLPE

The GRID program [20, 21] has been used by a number of authors [22, 23] as an alternative to the original CoMFA method for calculating the interaction fields. An advantage of the GRID approach, apart from the large number of chemical probes available, is the use of a 6–4 potential function, which is smoother than the 6–12 form of the Lennard-Jones type, for calculating the interaction energies at the grid lattice points. Good statistical results have been obtained, for example, in an analysis of glycogen phoshorylase b inhibitors by Cruciani et al. [24]. They used the GRID force field in combination with the GOLPE program [25], which performs the necessary chemometrical analysis. The dataset is particularly interesting because the X-ray structures of all protein-ligand complexes have been solved. This allowed the authors to investigate the dataset using new and different methods to further develop 3D QSAR techniques.

A further refinement of the original CoMFA technique has been achieved by introducing the concept of variable selection and reduction [3, 24]. As stated in section 2.6.1.3, the large number of variables in the descriptor matrix (i.e. the interaction energies) represents a statistical problem of the CoMFA approach. These variables make it increasingly difficult for multivariate projection methods, such as PLS, to distinguish the useful information contained in the descriptor matrix from information of a lesser quality or from noise. Thus, approaches for separating the useful variables from the less useful ones are needed. A statistical procedure named GOLPE (General Optimal Linear PLS Estimation) was developed by Baroni et al. [3] to achieve the objective of improving the predictivity of QSAR models. In the

GOLPE program several variable selection methods, such as D-optimal design and fractional factorial design (FFD), are implemented. The predictivity of each variable is determined by generating a large number of 3D QSAR models, and by calculating the SDEP. After completion of an FFD run, each variable is evaluated and classified into one of three categories: helpful for predictivity, detrimental for predictivity or uncertain. Only helpful variables are considered in the final PLS model. By applying the variable selection, models with higher cross-validated Q^2-values were usually derived compared to the corresponding conventional CoMFA models [11–13, 24, 26]. For a detailed description of this method see [3, 24, 26].

2.6.2.2 Alignment-independent Methods

The most crucial and difficult step in a CoMFA-related analysis is the realistic alignment of the moleculed studied. A further development of the CoMFA method that avoids the alignment problem has recently been described by several groups [27–29]. In their CoMMA (Comparative Molecular Moment Analysis) method Silverman et al. [27] used descriptors that characterize shape and charge distribution such as the principal moments of inertia and properties derived from dipole and quadrupole moments, respectively. The authors investigated a number of datasets and obtained models with good consistency and predictivity. A related approach using the GRID force field for the generation of principal moments has been reported by Cruciani et al. [28]. These authors have integrated their methods in the commercially available programs VOLSURF and ALMOND [30, 31]. For a detailed description of these relatively novel methods, the reader is referred to the literature [28, 29].

Although all CoMFA-related methods are of general use, a word of caution is necessary. There are a number of practical problems which emerge upon running the application. The results critically depend on the ligand conformation chosen, on the soundness of the alignment, on the chemical parameters used to describe the interaction fields and, last but not least, on the statistical evaluation method selected [14]. The reader should be aware of the fact that CoMFA is a powerful tool in the hand of the experienced user but may provide some difficulties for beginners.

2.6.3
More 3D QSAR Methods

Several other 3D QSAR approaches have been developed during the last few years. Some of them are not based on properties calculated within a lattice, as is the case for all CoMFA-related approaches. The GERM [32], COMPASS [33], receptor surface [34] and QUASAR [35] methods rely on properties calculated for discrete locations in the space at or near the union surface of the active ligands. The "receptor surface" thus generated is intended to simulate the macromolecular binding site. If all molecules of the data set bind in a manner that does not distort the residues at the binding site too much, this can be a workable approach. This approach is supported by the fact that reasonable models have been obtained by all of the aforementioned methods. However, two shortcomings remain associated with atomistic and receptor-surface models based on averaged receptor entities: the adaptation of the shape of the binding site by means of induced fit and hydrogen-bond adaptation. If the ligand-receptor interaction energy is determined with regard to an averaged receptor model, subtle effects associated with the adaptation of the receptor to the individual ligand molecules cannot be addressed. In addition, amino acid residues of a biological receptor with a conformationally flexible H-bond donor or acceptor can engage in differently directed H-bonds with different ligand molecules. This effect likewise cannot be simulated with an averaged receptor.

Another way to derive a quantitative structure-activity relationship is the generation of so called binding site or pseudoreceptor models. The pseudoreceptor modeling approach attempts to generate a 3D model of the binding site of a structurally unknown target protein based on the superimposed structures of known ligand molecules in their bioactive conformation, together with their experimentally determined binding affinities. The philosophy behind the "pseudoreceptor concept" is to engage the bound species in sufficient, specific non-covalent binding so as to mimic the essential interactions in the biological ligand-receptor complex (see, for example [37–40]). In the first step towards pseudoreceptor generation, potential binding sites (anchor points) are identified for each of the molecules of the pharmacophore. Suitable binding partners (e.g., amino acids, metal ions, water molecules) are then selected and positioned in three-dimensional space. After refinement, the ensemble of binding partners constitutes a pseudoreceptor for the ligand molecules that were used to generate it. In general, type and arrangement of the pseudoreceptor building blocks surrounding the pharmacophore model will bear no structural resemblance to the real biological target protein. Instead of reproducing the complex structure of the ligand-binding protein, the receptor surrogate should be regarded as a purely hypothetical model of the binding pocket, accommodating a series of structurally related ligands. The estimation of binding affinities relies on the evaluation of ligand-pseudoreceptor interaction energies, ligand desolvation energies and changes in ligand internal energy and entropy upon receptor binding. The pseudoreceptor concept, integrated in the PrGen program [41], has been validated by constructing receptor surrogates for the enzyme human carbonic anhydrase, the dopaminergic and the β_2-adrenergic receptor. The predicted free energies of ligand binding in the pseudoreceptor and experimental values derived from the biological receptor agree to within 1.2 kcal/mol [42]. Advantages of the pseudoreceptor concept are the use of a directional force field that is capable of correctly treating hydrogen bonds and ligand-metal ion-protein interactions, frequently found to be of prime importance for the binding of drug molecules, and its consideration of solvation and entropic processes, often missing in 3D QSAR approaches.

References

[1] Cramer, R. *J. Comput.-Aided Mol. Design* **6**, 475–486 (1992).

[2] Klebe, G., Abraham, U. and Mietzner, T. *J. Med. Chem.* **37**, 4130–4146 (1994).

[3] Baroni, M., Constantino, G., Cruciani, G., Riganelli, D., Valigi, R., and Clementi, S. *Quant. Struct.-Act. Relat.* **12**, 9–20 (1993).

[4] Martin, Y. C. *Perspectives in Drug Discovery and Design* **12**, 3–23 (1998).

[5] Klebe, G. and Abraham, U. *J. Med. Chem.* **36**, 70–80 (1993).

[6] Kim, K. H., Greco, G. and Novellino, E. *Perspectives in Drug Discovery and Design* **12**, 257–315 (1998).

[7] Folkers, G., Merz, A. and Rognan, D. *CoMFA: Scope and limitations.* In: Kubinyi, H. (Ed.) 3D QSAR in drug design: Theory, methods and applications, ESCOM, Leiden, The Netherlands, 1993, pp. 583–618.

[8] Wold, S. *Quant. Struct.-Act. Relat.* **10**, 191–193 (1991).

[9] Wold, S., Johansson, E. and Cocchi, M. *PLS – Partial least squares projections to latent structures.* In: Kubinyi, H. (Ed.) 3D QSAR in drug design: Theory, methods and applications, ESCOM, Leiden, The Netherlands, 1993, pp. 523–550.

[10] Krystek, S. R., Hunt, J. T., Stein, P. D. and Stouch, T. R. *J. Med. Chem.* **38**, 659–668 (1995).

[11] Sippl, W. *J. Comput. Aided Mol. Design* **14**, 559–572 (2000).

[12] Sippl, W., Contreras, J.M., Parrot, I., Rival, Y.M. and Wermuth, C.G. *J. Comput.-Aided Mol. Des.* **15**, 395–410 (2001).

[13] Sippl, W. *Bioorg. Med. Chem.* **10**, 3741–3755 (2002).

[14] Oprea, T. I. and Garcia, A. E. *J. Comput. Aided Mol. Des.* **10**, 186–200 (1996).

[15] Norinder, U. *Perspectives in Drug Discovery and Design* **12**, 25–39 (1998).

[16] Oprea, T. I., Waller, C. L. and Marshall, G. R. *J. Med. Chem.* **37**, 2206–2215 (1994).

[17] Klebe, G. and Abraham, U. *J. Comput. Aided Mol. Design* **13**, 1–10 (1999).

[18] Böhm, M., Stürzebecher, J. and Klebe, G. *J. Med. Chem.* **42**, 458–477 (1999).

[19] Böhm, M. and Klebe, G. *J. Med. Chem.* **45**, 1585–1597 (2002).

[20] Goodford, P. J. *J. Med. Chem.* **28**, 849–857 (1985).

[21] Wade, R. C., Clark, K. J., and Goodford, P. J. *J. Med. Chem.* **36**, 140–147 (1993).

[22] Davis, A. M., Gensmantel, N. P., Johansson, E. and Marriot, D. P. *J. Med. Chem.* **37**, 963–972 (1994).

[23] Kim, K. H., Greco, G., Novellino, E., Silipo, C. and Vittoria, A. *J. Comput.-Aided Design* **7**, 263–280 (1993).

[24] Cruciani, G. and Watson, K. A. *J. Med. Chem.* **37**, 2589–2601 (1994).

[25] GOLPE, Multivariate infometric analysis, Perugia, Italy.

[26] Cruciani, G., Clementi, S. and Pastor, M. *Perspectives in Drug Discovery and Design* **12**, 71–86 (1998).

[27] Silverman, B. D. and Platt, D. E. *J. Med. Chem.* **39**, 2129–2140 (1996).

[28] Cruciani, G., Crivori, P. Carupt, P. A. and Testa, B. *Theochem* **503**, 17–31 (2000).

[29] Pastor, M., Cruciani, G., McLay, I., Pickett, S., Clementi, S. *J. Med. Chem.* **43**, 3233–3243 (2000).

[30] VOLSURF, Molecular Discovery Ltd., Oxford, UK.

[31] ALMOND, Multivariate infometric analysis, Perugia, Italy.

[32] Walters, E. *Perspectives in Drug Discovery and Design* **12**, 159–166 (1998).

[33] Jain, A. N., Koile, K. and Chapman, D. *J. Med. Chem.* **37**, 2315–2327 (1994).

[34] Hahn, M. and Rogers, D. *Perspectives in Drug Discovery and Design* **12**, 117–134 (1998).

[35] Vedani, A. and Zbinden, P. *Pharm. Acta Helv.* **73**, 11–18 (1998).

[36] Vedani, A., Zbinden, P. and Snyder, J. P. *J. Recept. Res.* **13**, 163–177 (1993).

[37] Sippl, W., Stark, H. and Höltje, H.-D. *Pharmazie* **53**, 433–437 (1998).

[38] Höltje, H.-D. and Jendretzki, U. K. *Arch. Pharm.* **328**, 577–584 (1995).

[39] Greenidge, P. A., Merz, A. and Folkers, G. *J. Comput.-Aided Mol. Design* **9**, 473–478 (1995).

[40] Schmetzer, S., Greenidge, P. A., Kovar, K. A., Schulze-Alexandru, M. and Folkers, G. *J. Comput.-Aided Mol. Design* **11**, 278–292 (1997).

[41] PrGen, Biographics Laboratory, Basel, Switzerland.

[42] Vedani, A., Zbinden, P., Snyder, J. P. and Greenidge, P. A. *J. Am. Chem. Soc.* **117**, 4987–4994 (1995).

3
A Case Study for Small Molecule Modeling:
Dopamine D_3 Receptor Antagonists

In this chapter we describe the determination of a pharmacophore model and a sub-sequent 3D QSAR analysis based on the GRID/GOLPE method (section 2.6.2.2) for dopamine D_3 receptor antagonists. We will use the steric and electrostatic informa-tion derived from partially rigidized, highly affine ligands to build up the pharmaco-phore model. After definition of the arrangement of the pharmacophoric descrip-tors, the model will be validated by closer examination of molecular fields that are generated by the ligands superimposed onto each other in their pharmacophoric conformations. In a last step, the molecular fields that are produced by the program GRID are correlated with the binding affinities by means of a PLS model that is drawn up by program GOLPE. This PLS model is validated using different cross-validation techniques and its predictive power is tested by an external test set of ligands.

For building the pharmacophore model, sterically restricted dopamine D_3 recep-tor antagonists from the laboratory of Prof. H. Stark, University of Frankfurt/Main, Germany [1], and antagonists described in the literature were used.

40 ligands are used for the subsequent GRID/GOLPE analysis (Tables 3.1, 3.2 and 3.3). All of these ligands were obtained from our co-worker's laboratory to assure consistency in the binding data.

3.1
A Pharmacophore Model for Dopamine D_3 Receptor Antagonists

Five ligands that bind as antagonists to the dopamine D_3 receptor have been exam-ined in detail to define their bioactive conformation (Table 3.4). The molecular struc-ture of the antagonists that are included in this analysis can be regarded as a compo-sition of three fragments. They consist of an aromatic-basic element, an aromatic-amidic portion, and an aromatic or alkylic spacer. Fig. 3.1 shows the decomposition of substance BP897 [2] into the aforementioned fragments. Variations of this struc-ture are shown together with their binding affinities in Table 3.1.

Most of the ligands described as dopamine D_3 antagonists fit into this scheme. However, there are some ligands, e. g. compound 1 in Table 3.1, which have their amidic portions replaced by other moieties that are able to accept hydrogen bonds.

Tab. 3.1 Dopamine D_3 antagonists with variations in their amidic portion and in the length of the spacer.

ST-	R1	R2	n	D_3 K_i (nm)	ST-	R1	R2	n	D_3 K_i (nm)
63	*(structure)*	H	4	26	64	*(structure)*	H	4	43
65	*(structure)*	H	4	10	66	*(structure)*	OCH₃	3	213
67	*(structure)*	OCH₃	3	300	68	*(structure)*	OCH₃	3	396
69	*(structure)*	OCH₃	4	7.8	70	*(structure)*	OCH₃	4	9.2
71	*(structure)*	OCH₃	4	3.9	82	*(structure)*	OCH₃	4	2.85
84	*(structure)*	OCH₃	4	38	85	*(structure)*	OCH₃	4	23.3
86	*(structure)*	OCH₃	4	29	88	*(structure)*	OCH₃	3	560
92	*(structure)*	OCH₃	4	32	93	*(structure)*	OCH₃	3	108
95	*(structure)*	OCH₃	4	2.5	96	*(structure)*	OCH₃	4	1
98	*(structure)*	OCH₃	4	6.6	99	*(structure)*	OCH₃	4	1.48
100	*(structure)*	OCH₃	4	9.62	101	*(structure)*	OCH₃	4	2.96
144	*(structure)*	OCH₃	4	1.8	150	*(structure)*	OCH₃	4	42
152	*(structure)*	OCH₃	4	6.3	167	*(structure)*	OCH₃	4	0.50
168	*(structure)*	OCH₃	4	0.61	188	*(structure)*	OCH₃	4	0.69
189	*(structure)*	OCH₃	4	2.14	317	*(structure)*	OCH₃	4	10.2

Tab. 3.2 Dopamine D_3 antagonists with variations in their spacer.

ST-	X	D_3 K_i (nm)	ST-	X	D_3 K_i (nm)
81		40	176		100
177		750	205		37.2

Tab. 3.3 Dopamine D_3 antagonists with aminotetralines in their aromatic-basic portion.

ST-	R	D_3 K_i (nm)	ST-	R	D_3 K_i (nm)
124		233	125		48
126		28	127		60.5
185		38			

Tab. 3.4 Dopamine D_3 antagonists used for the definition of the pharmacophore.

Compound	Structure	D_3 K_i (nM)
1 [3]		28
2 [4]		1
ST-205		37.2
ST-84		38
ST-85		23.3

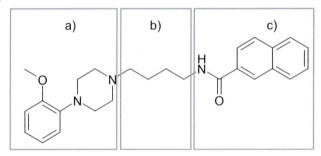

Fig. 3.1 BP 897 decomposed in three fragments. a) The aromatic-basic fragment, b) the spacer, and c) the aromatic-amidic fragment.

Fig. 3.2 Blueprint for dopamine D₃ receptor antagonists.

Therefore, all ligands can roughly be described by the representation shown in Fig. 3.2. As all ligands examined contain the same pharmacophoric descriptors (e.g. basic nitrogens, aromatic moieties, H-bond acceptors) we may assume that they bind in a very similar mode to the same binding pocket of the dopamine D₃ receptor. To detect the bioactive conformation these ligands adopt in the binding pocket, we concentrated on the analysis of the conformational space of those ligands that show rigidisation in parts of their structures.

Since none of the ligands examined is completely rigidized, but some are partially rigidized, those molecules were in a first step decomposed into three fragments and the conformationally restricted fragments were then examined individually. The fragments chosen were overlapping to some extent in order to also determine the bioactive conformation of the parts linking them. After the putative bioactive conformations for the fragments were defined, the structures were reassembled yielding the putative binding conformations.

3.1.1
The Aromatic-Basic Fragment

Most of the structures in our dataset (Tables 3.1, 3.2 and 3.3) contain a N-(4-(2-methoxyphenyl)piperazine-1-yl) moiety which is rather flexible, since various energetically favourable conformations can be adopted by the ring system. In contrast, ligand compound 2 is rigidized in that part of the structure because it contains a ring system without any rotatable bonds instead.

In a first step, therefore, the conformational space of the octahydrobenzoquinoline ring system of compound 2 was examined in detail. For this purpose a simulated annealing procedure (see section 2.3.3) was performed. The 4-methyl-1,2,3,4,4a,5,6,10b-octahydrobenzo[f]quinoline-7-ol fragment was heated up ten times to 2000 K and was subsequently annealed to 0 K. The low temperature conformations were collected and inspected visually. Two clusters of very similar low energy conformations exhibiting the possible conformations of this ring system were found. Representatives of these clusters are shown in Fig. 3.3.

The 1-(2-methoxyphenyl)-4-methylpiperazine ring system was fitted onto both structures shown in Fig. 3.4 using the FlexS [5] program (see section 2.5.2). FlexS not only includes the sterical and electronical demands of the fragments in the superposition procedure, but also proposes virtual interacting points that may serve as counter ions or hydrogen bond partners. In Fig. 3.4 the superposition of the phenylpiperazine ring system onto both conformations of the tricyclic system of compound 2 is depicted. The conformation of the phenylpiperazine is the same in both cases. As can be seen, the different fragments have the possibility to interact with the same putative hydrogen bond donors and salt bridge partners.

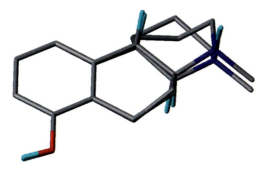

Fig. 3.3 Representative structures of the octahydrobenzoquinoline ring system.

Fig. 3.4 Superposition of 1-(2-methoxyphenyl)-4-methylpiperazine (white carbon atoms) onto both low energy conformers of 4-methyl-1,2,3,4,4a,5,6,10b-octahydrobenzo[f]quinoline-7-ol (grey carbon atoms) by FlexS. Virtual interacting points are shown as orange balls.

3.1.2
The Spacer

The fragments representing the spacer are the most flexible parts of the ligands in the dataset. It is, therefore, difficult to decide which conformation of this part is the one that binds to the receptor. Fortunately, the corresponding spacer fragments of compound 1 and ST-205 (Fig. 3.5) are at least rigidized to some extent and were thus examined in detail.

In a first step, the conformational space of the ST-205 spacer was determined. The bicyclic ring system was subjected to a simulated annealing which lead to the three different conformations displayed in Fig 3.6.

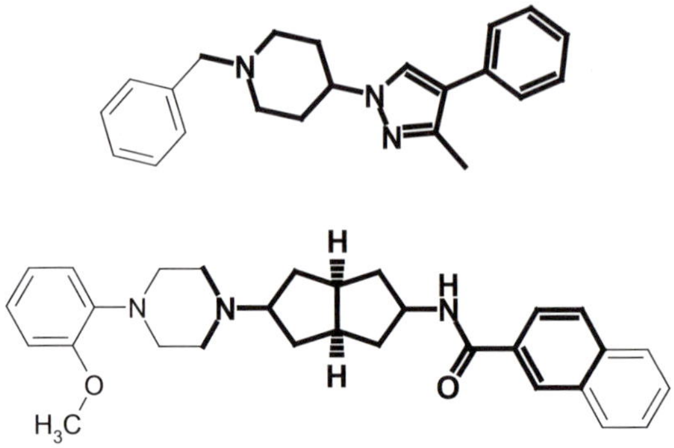

Fig. 3.5 Compounds 1 and ST-205. Fragments that were examined in detail to define the conformation of the spacer are displayed in bold.

Fig. 3.6 Fragment of ST-205. Three conformations are possible for the ring system.

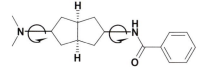

Fig. 3.7 Extended spacer of ST-205. The indicated bonds are rotated in steps of 10°.

Fig. 3.8 Fragments of compound 1 and ST-205. The superposition that was rated best by FlexS is shown.

The ring system was then extended to include parts of the neighbouring fragments and the rotatable bonds of the extended fragment were examined by a systematic search (Fig. 3.7).

The simulated annealing procedure of the bicyclus in combination with the systematic search lead to 992 possible conformations of the extended spacer fragment of ST-205. The analogous fragment of compound 1 was then flexibly fitted onto each of the 992 possible conformations. Each superposition was rated by a scoring function implemented in FlexS. The superposition with the highest rating was assumed to be the binding conformation of these fragments. This conformation is shown in Fig. 3.8.

3.1.3
The Aromatic-Amidic Residue

In several of the compounds which bind with high affinity to the dopamine D$_3$ receptor, the so called 'aromatic-amidic fragment' is represented by phthalimide moieties. It is easy, therefore, to determine the pharmacophoric conformation for this fragment of the ligands.

Fig. 3.9 shows the phthalimide residues present in compounds ST-84 and ST-85. The planar conformation depicted in Fig. 3.9 is the only possible low energy conformation for these structural elements.

3.1.4
Resulting Pharmacophore

After determination of the preferred conformation of all fragments the ligands were reassembled. Fig. 3.10 shows four ligands superimposed onto each other in their

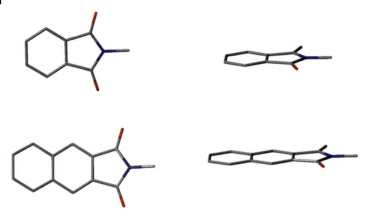

Fig. 3.9 Planar phthalimide structures that can be found in ligands ST-84 and ST-85.

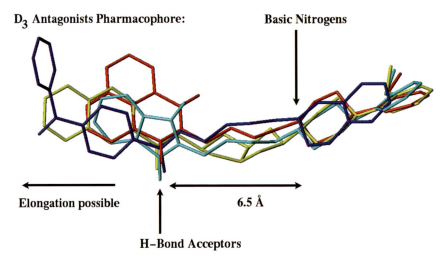

Fig. 3.10 Pharmacophore for dopamine D₃ receptor antagonists. Four ligands are shown in their pharmacophoric conformation. ST-127 (blue), ST-205 (yellow), ST-84 (cyan) and ST-86 (red).

putative bioactive conformations. The ligands adopt a stretched conformation. The basic nitrogens, the H-bond acceptors and the lipophilic residues occupy distinct areas. The pharmacophoric points that can form directional interactions with the receptor, i. e. the H-bond acceptors and the basic nitrogens, are about 6.5 Å apart from each other. The aromatic region at the aromatic-amidic part of the ligand can be considerably elongated.

3.1.5
Molecular Interaction Fields

As mentioned in section 2.5.3, molecules recognize each other by characteristic properties which can extend beyond their van der Waals volume. Therefore, pharmacophoric superpositions should not only be estimated by the consideration of super-imposed atomic skeletons, but also should take into account the interaction fields generated by them.

Fig. 3.11 shows four high affinity ligands in their pharmacophoric conformations superimposed onto each other. Molecular interaction fields for each ligand are calculated by the GRID program [6] using different probes. The basic nitrogens of the ligands are protonated for this analysis corresponding to the situation under physiological conditions. As it can be seen in Fig. 3.11 b, c and d, the protonated nitrogens of all ligands in a similar way interact favourably with the 'ionized alkyl carboxyl group'-probe (Fig. 3.11 c). All ligands also interact with the 'sp^2 NH with lone pair'-probe (Fig. 3.11 b) and show favourable lipophilic interactions with the 'sp^2 CH'-probe (Fig. 3.11 d). In all cases the corresponding fields occupy similar regions. The described interactions seem to be crucial for the binding to the receptor. Some other favourable ligand-probe interactions are only formed by some of the members of the series. This behaviour is demonstrated in Fig. 3.11 a which shows the interaction of ligands with the hydrogen bond accepting 'carbonyl oxygen'-probe. Only ligands ST-127 and ST-205 are able to donate hydrogen bonds in their amidic portion and thus contribute favourable interaction energies in this area. However, since all ligands displayed in Fig. 3.11 bind with high affinity to the receptor, this ability seems to be of less importance. The information derived from the GRID fields can be used to develop ideas for the binding site of the receptor. In the case under study, an amino acid that can donate a hydrogen bond, another one that forms a salt-bridge and several additional lipophilic amino acids are most likely to constitute the interacting

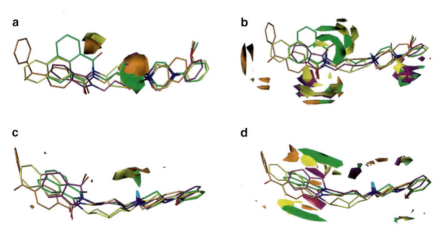

Fig. 3.11 Four ligands ST-205 (yellow carbons), ST-84 (violet carbons), ST-127 (orange carbons) and ST-86 (green carbons) are displayed with their GRID contours in corresponding colours.

a) GRID contours derived from the 'carbonyl oxygen'-probe contoured at –2,5 kcal mol^{-1}.

b) GRID contours derived from the 'sp^2 NH with lone pair'-probe. Energy contoured at –4 kcal mol^{-1}. c) GRID contours derived from the 'ionized alkyl carboxyl group'-probe contoured at –3.5 kcal mol^{-1}. d) GRID contours derived from the aromatic 'sp^2 CH'-probe contoured at –1.2 kcal mol^{-1}.

partners of D_3 receptor antagonists. The probable relative spatial arrangement of these amino acids is defined by the corresponding GRID-fields.

3.2
3D QSAR Analysis

40 D_3 antagonists (Tables 1, 2 and 3) were superimposed onto each other in their pharmacophoric conformations. The superposition was conducted using FLEXS and Sybyl's Multifit routine for the refinement of the superpositions. The resulting superimposition of all 40 ligands is shown in Fig. 3.12. GRID interaction fields are calculated subsequently using the phenolic hydroxyl group probe (OH) which is placed on each node of a grid box surrounding the ligands. The size of the grid box was defined so that it extends about 4 Å from the structures of the ligands. GRID calculations were carried out using 1 Å grid spacing, thus resulting in 14580 probe-ligand interactions for each compound.

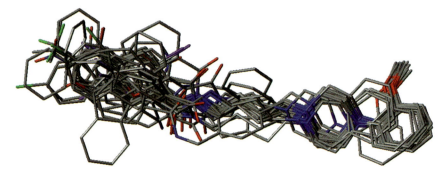

Fig. 3.12 All dopamine D_3 antagonists superimposed in the pharmacophoric conformation.

3.2.1
Variable Reduction and PLS model

As stated in section 2.6.2.2 the large number of variables (i.e. interaction energies) in the descriptor matrix presents a statistical problem for multivariate projection methods. Only some of the calculated interaction energies contain useful information whereas others only introduce noise into the statistical analysis. Therefore the program GOLPE [7] has been used to perform a variables selection and set up a PLS model.

Starting with all 14580 variables, the data was pretreated in a way that those variables were removed from the model which only assume two or three distinct values as well as those with absolute values below 10^{-7} kcal/mol. This first step led to a reduction to 13665 variables that were considered further. Next, the variables were classified using a principal component analysis (PCA) and a first PLS model was generated and cross-validated using the „leave one out" method (LOO). We continued to reduce the variables several times using the D-optimal preselection method implemented in the program and obtained a PLS-model built up from 1682 variables. This model was equivalent in quality compared to the first one, when it was checked using the LOO cross validation method. At this step the fractional factorial design method (FFD) combined with the smart region definition (SRD) [8] was chosen to reduce the number of variables for a last time and to generate the final model. As mentioned in section 2.6.2.2 this variable selection assorts very carefully those variables contributing to the predictivity of the model. The final model was validated using the LOO and the „leave 20% out" method.

$$SDEP = \sqrt{\sum \frac{(Y-Y')^2}{N}} \tag{1}$$

$$R^2 = 1 - \frac{\sum(Y-Y')^2}{\sum(Y-\bar{Y}')^2} \tag{2}$$

Y: experimental value, Y':calculated value, Ȳ': average value, N: number of objects

The results of the 3D QSAR analysis are shown in table 3.5. Three principal components have been used to generate the models. The characteristics used to describe the quality of the models are the number of variables, the R^2 values (correlation coef-

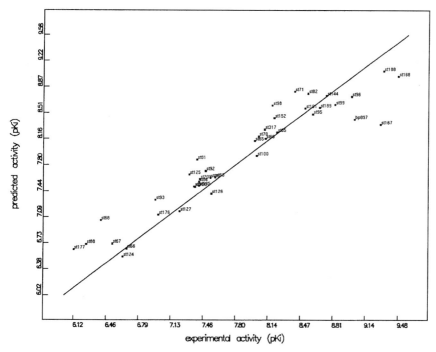

Fig. 3.13 Cross validated PLS model (leave one out).

Tab. 3.5 PLS models generated with variables from three principal components.

# Variables	Selection method	R^2	SDEC	Q^2	SDEP	Validation method
13665	–	0.9545	0.1863	0.7041	0.4753	LOO
6728	D-optimal	0.9545	0.1863	0.7041	0.4753	LOO
3364	D-optimal	0.9545	0.1863	0.7041	0.4753	LOO
1682	D-optimal	0.9545	0.1683	0.7044	0.4751	LOO
799	SRD/FFD	0.9673	0.1580	0.8743	0.3098	LOO
799	SRD/FFD	0.9673	0.1580	0.8549	0.3328	L 20% O

ficients, Eq. 2), the SDEC (standard deviation of the error of correction, Eq. 1), the Q^2 values (cross validated correlation coefficient, Eq. 1 in section 2.6.1.3), and the SDEP (standard deviation of the error of prediction, Eq. 2 in section 2.6.1.3). Fig. 3.13 displays the result of the 'leave one out' cross validation of the final model. The correlation of the predicted and the experimental determined pK$_i$ values is shown.

3.2.2
Validation of the Method

In a next step the 3D QSAR method itself has been validated. As mentioned in section 2.6, the treatment of a vast amount of data by application of statistical methods can result in a chance correlation. The underlying dataset has been tested using a scramble test to check whether the obtained correlation has been generated by chance or can be considered as reasonable. The binding affinities of the ligands were scrambled and randomly assigned to the ligands. A PLS model was generated and the variables were reduced as mentioned above. The final PLS model was checked using the LOO cross validation method. Ten models with scrambled affinities were set up. The characteristics of the resulting models are shown in table 3.6.

Tab. 3.6 Models obtained after scrambling binding affinities.

Model	R^2	SDEC	Q^2	SDEP
1	0.7449	0.4413	−0.4612	1.0562
2	0.7871	0.4032	0.1757	0.7923
3	0.7874	0.4028	0.2367	0.7634
4	0.8521	0.3360	0.3356	0.7122
5	0.8066	0.3843	0.2233	0.7700
6	0.8719	0.3129	0.3979	0.6780
7	0.7481	0.4385	−0.4564	1.0545
8	0.8176	0.3732	−0.2714	0.9852
9	0.9128	0.2581	−0.1091	0.9202
10	0.8010	0.3898	−0.0810	0.9085

Interestingly, rather high correlation coefficient values (R^2) were obtained by each PLS model, but none of the models survived the cross validation, which is indicated by the Q^2 values that span a range from –0.4564 to 0.3979. The SDEC and SDEP values in all cases are also very high compared to the model with correctly assigned pK_i values. These results clearly show that the method is able to generate good models only if binding affinity values have been correctly assigned.

3.2.3
Prediction of External Ligands

In a last step the predictivity of the model was tested on a set of ligands which were not included in the model (table 3.7). The binding affinities of 12 ligands that were synthesized and tested in the same laboratories as the training set were predicted and compared to their actual pK_i-values. The structures of these ligands differ from the ones treated so far in this chapter. Therefore, the prediction of their binding affinities is even more challenging. Unfortunately, the structures of the ligands cannot be shown due to patent protection.

The SDEP value for the external prediction is 0.57. The predicted pK_i values in most cases fall within the range of + 0.5 compared to the experimentally determined values, what must be rated as a reasonable result.

Tab. 3.7 Prediction of binding affinities of an external test set of ligands.

Compound	Experimental pK_i	Predicted pK_i	Compound	Experimental pK_i	Predicted pK_i
ST-73	6.62	7.33	ST-106	7.12	7.57
ST-75	6.66	7.10	ST-109	6.97	7.70
ST-76	6.71	7.32	ST-111	6.75	7.32
ST-78	6.39	7.02	ST-115	7.13	7.59
ST-87	7.62	7.91	ST-128	7.66	7.50
ST-104	8.69	8.04	ST-129	8.55	8.10

References

[1] Hackling, A., Ghosh, R., Perachon, S., Mann, A., Höltje, H.-D., Wermuth, C. G., Schwartz, J.-C., Sippl, W., Sokoloff, P., and Stark, H. *J. Med. Chem.* (in press).

[2] Pilla, M., Perachon, S., Sautel, F., Garrido, F., Mann, A., Wermuth, C. G., Schwartz, J.-C., Everitt, B. J., and Sokoloff, P. *Nature* **400**, 371–375 (1999).

[3] Moore, K. W., Bonner, K., Jones, E. A., Emms, F., Leeson, P. D., Marwood, R., Patel, S., Rowley, M., Thomas, S., and Carling, R. W. *Bioorg. Med. Chem. Lett.* **9**, 1285–1290, (1996).

[4] Avenell, K. Y., Boyfield, I., Coldwell, M. C., Hadley, M. S., Healy, M. A. M., Jeffrey, P. M.,

Johnson, C. N., Nash, D. J., Riley, G. J., Scott, E. E., Smith, S. A., Stacey, R., Stemp, G., and Thewlis, K. M. *Bioorg. Med. Chem. Lett.* **8**, 2859–2864 (1998).

[5] Lemmen, C., Lengauer, T., and Klebe, G. *J. Med. Chem.* **41**, 4502–4520 (1998).

[6] GRID 19, Molecular Discovery Ltd., London, UK.

[7] GOLPE, Multivariate infometric analysis, Perugia, Italy.

[8] Cruciani, G., Clementi, S., and Pastor, M. *Perspectives in Drug Discovery and Design*, **12/13/14**, 71–86 (1998).

4
Introduction to Comparative Protein Modeling

4.1
Where and How to get Information on Proteins

Within this book we have, until now, been discussing small molecules. In the second part of the book, the topic of discussion will be *biopolymers*. Since most of the receptors and target molecules known are polypeptides, the main part of the discussion will center on the modeling of proteins.

Each modeling study depends heavily on the quality of the available experimental data, which always serve as the basis of a hypothetical model. Therefore, the first step should always be a careful literature and database search in order to get a clear picture about the level of knowledge on the biopolymer structure of interest. Valuable information would for example be the complete 3D structure of the receptor or enzyme, ideally derived from crystal data or NMR measurements. After refinement such a structure can be used directly to calculate different properties of the protein or to investigate possible ligand–protein interactions.

Despite the rapid growth of the number of solved 3D structures of proteins, the rate of increase of sequence data continues to be greater, resulting in an even larger number of protein sequences for which no 3D structure is known.

Since the beginning of the 1990s, many laboratories are analyzing the full genome of several species of bacteria, of yeasts, mice, and humans. During these collaborative efforts enormous amounts of sequence data are collected and stored in databases, most of which are publically accessible. In addition to gathering all these data, it is more than ever necessary to compare the nucleotide or amino acid sequences and search for similarities and differences.

Since the number of published sequences and structural information is increasing rapidly an efficient search can only be done by using computer software suitable for this purpose. Several algorithms have been developed that provide graphical user interfaces to existing databases. Thus, a comparison of a newly found sequence with those already stored in a database can nowadays be performed in a few minutes. Nevertheless, it is still necessary to carefully analyze the results and to fine-tune a data base search if needed. Following this advice, it is possible to quickly determine the sequence differences among several species and between a healthy versus a diseased individual.

One such well-known system is the GCG program [1] offered by the Genetic Computer Group, Wisconsin and now implemented in the Accelrys software [2]. This package allows work with several databases which can be used for the search of an individual protein or DNA structure. The search can be accelerated and specified by employing keywords like author names, journals or families of proteins.

A variety of nucleotide and protein sequence databases are maintained by the scientific community and are available via the World Wide Web. The EMBL Nucleotide Sequence Database [3] constitutes Europe's primary nucleotide sequence resource. Its main sources for DNA and RNA sequences are direct submissions from individual researchers, genome sequencing projects and patent applications. The TrEMBL database [4] contains the translations of all coding sequences present in the EMBL Nucleotide Sequence Database, which are not yet integrated into the SWISS-PROT database [5]. SWISS-PROT is a scientifically managed protein sequence database that provides a high level of annotation, a minimal level of redundancy and high level of integration with other databases. It is maintained collaboratively by the Swiss Institute for Bioinformatics and the European Bioinformatics Institute. SWISS-PROT offers a high level of information including a description of the function of a protein, the structure of its domains, etc. The PIR database [6] which is related to SWISS-PROT contains about 200,000 protein sequences, which are classified into families and superfamilies. Their domains and motifs are identified. The PIR web site (http://pir.georgetown.edu) features search engines that use sequence similarity and database annotation to facilitate the analysis and functional characterization of proteins.

The most important and standard database for all structural information on macromolecules is the Protein Data Bank [7] which is available via the World Wide Web (http://www.rcsb.org). In the Protein Data Bank atomic coordinates of protein or DNA structures are collected. Because of the continuously growing number of experimentally resolved structures the database is regularly updated. Information hunting in the Protein Data Bank can be performed by specifying particular keywords: the author name, a journal, or a part of a sequence for example can serve as a search subject.

Based on the Protein Data Bank some smaller structural databases have been created. Examples are the HSSP [8] and the SCOP [9] databases. HSSP contains homology-derived structures of proteins. It combines information from the Protein Data Bank and sequences of proteins derived from a sequence database like SWISS-PROT. The SCOP database is a comprehensive alignment of all proteins of known structure, according to their evolutionary and structural relationships. Protein domains in SCOP are grouped into species and hierarchically classified into families, superfamilies, folds and classes.

In general the format, organization and information contained in different structural data files is very similar. As the Protein Data Bank is widely used the standard format of a Protein data file (PDB) will be described in detail in the following. The header of the data file comprises some general information about the protein. It includes the official name, references, resolution of the crystal structure and some useful remarks about the secondary structure composition of the protein. Adjacent to the header are listed the atomic coordinates. Atoms belonging to standard amino acid residues are labelled as ATOM. In order to distinguish between individual peptide chains the ATOMS are separated by an additional line starting with the abbreviation TER. Between ATOMS a bond is generally built when the file is read into the modeling program. This is important as the atoms which do not belong to standard amino acid residues are labelled as HETATM. No connectivity is established between HETATMS. Therefore an additional connectivity table is included at the end of the data file. It is advisable to be careful at this point because it is program-dependent whether or not HETATMS are displayed properly and connectivities are correctly assigned.

HETATMS can either belong to non-standard amino acids or, in the case of complexes, to the ligand molecule involved in the ligand–protein interaction. As the proposed atom type assignment is often incorrect it is absolutely necessary to check carefully all atom types to avoid mistakes resulting in wrong geometries of the ligands (this has already been discussed in Chap. 2.1.2).

Usually, all structures from the Protein Data Bank do not include hydrogen atoms. For some types of investigation hydrogen atoms can be neglected but for the study of ligand–protein interactions it is inevitable to add the hydrogens. The ligand molecules have to be checked especially carefully in order to confirm that the correct degree of protonation has been assigned in the case of acidic or basic substances.

In addition hydrogen atoms never are allocated to all water molecules. As a consequence they are displayed only as single points representing the oxygen positions. Water molecules can present either simple crystal water distributed near the surface of the protein, or they can be located in the active site. In the latter case it is absolutely necessary to include their complete coordinates into further investigations because they can crucially influence the conformation of the active site structure. This is also true for cations implemented in the crystal structure as they can play an important role for ligand binding or enzyme activity if they are located in the active site.

Most of the modeling programs are able to read the PDB files without problems and to transform the structural information into a 3D picture of the protein. However, some points of caution should be kept in mind when using experimentally derived information.

In principle, the resolution of a crystal structure should be at least between 2.5 Å and 1.5 Å or better, otherwise the structural information is not very valuable. The purification process of proteins is a difficult and time-consuming task and it may happen that as a result of proteolytic activity some information could be lost before the crystallization process has finished. Therefore amino residues may sometimes be missing, leading to incomplete information contained in the data file.

Some enzymes and proteins fulfil their biochemical function only in the dimeric or trimeric form. The modeler should be aware of this fact because it makes no sense to investigate the functionality of the active site of an enzyme which consists of a dimer when only the monomer structure is present in the PDB file.

Recently the NMR technique has become a frequently used method for obtaining structural information on proteins. NMR has a special bearing on cases where a protein has withstood all efforts to grow sufficiently large crystals. An additional advantage of NMR-derived data is that the conformation of the protein is not influenced by packing forces of the crystal environment. As the NMR measurements are performed in solution the results are highly dependent on the solvent. Experiments in apolar solvents for example lead to an overestimation of hydrogen bonding phenomena. Measurements in aqueous environment should yield a more realistic picture of the protein structure.

The pool of information on proteins is already immense and is growing continuously. However, most of the available databases still only contain information on primary structures. In order to obtain a 3D protein model from these data the application of alignment techniques, knowledge-based and comparative modeling approaches is necessary. A detailed discussion on these subjects will be given in section 4.3.

References

[1] Devereux, J., Haeberli, P., and Smithies, O. *Nucleic Acids Res.* **12**, 387–395 (1984).

[2] GCG, Accelrys Inc., San Diego, USA

[3] Emmert, D. B., Stoehr, P. J., Stoesser, G., and Cameron, G. N. *Nucleic Acids Res.* **22**, 3445–3449 (1994).

[4] O'Donovan, C., Martin, M.J., Gattiker, A., Gasteiger, E., Bairoch, A., and Apweiler, R. *Brief Bioinform.* **3**, 275–284 (2002).

[5] Bairoch, A., and Boeckmann, B. *Nucleic Acids Res.* **22**, 3578–3580 (1994).

[6] George, D. G., Barker, W. C., Mewes, H.-W., Pfeiffer, F., and Tsugita, A. *Nucleic Acids Res.* **22**, 3569–3573 (1994).

[7] Bernstein, F. C., Koetzle, T. F., Williams, G. J. B., Meyer, E. F., Brice, M. D., Rodgers, J. R., Kennard, O., Shimanouchi, T., and Tasumi, M. *J. Mol. Biol.* **112**, 535–542 (1977).

[8] Sander, C., and Schneider, R. *Nucleic Acids Res.* **22**, 3597–3599 (1994).

[9] Lo Conte, L., Brenner, S. E., Hubbard, T. J. P., Chothia, C. and Murzin, A. G. *Nucleic Acids Res.* **30**, 264–267 (2002).

4.2
Terminology and Principles of Protein Structure

The complex 3D structure of proteins can be characterized in four general levels of structural organization: primary, secondary, tertiary and quaternary structure.

1. The primary structure represents the linear arrangement of the individual amino acids in the protein sequence.
2. The secondary structure describes the local architecture of linear segments of the polypeptide chain (i.e. α-helix, β-sheet), without regarding the conformation of the side chains. Another level of structural organization, which was introduced not before very recently, is the so-called *supersecondary structure*. It describes the association of secondary structural elements through side chain interactions. Another term for the same matter is "motif".
3. The tertiary structure portrays the overall topology of the folded polypeptide chain.
4. The quaternary structure describes the arrangement of separate subunits or monomers into the functional protein.

Owing to the remarkable capability of polypeptide chains not only in vivo but also in vitro to fold into functional proteins, it is currently accepted that most aspects of protein architecture and stabilization directly derive from the properties of the particular sequence of amino acids that make up the polypeptide chain (i.e. the primary structure). These properties include the individual characteristics of the side chains of every residue and the influence of the polypeptide backbone on the protein conformation. Only on the basis of this information can 3D structure of a protein be understood. It is not the scope of this introduction to provide a detailed description of all the properties which determine the conformation of a protein, but to explain the main features necessary to understand the contents of the following sections. For a comprehensive description of the principles of protein conformation, the reader is referred to the literature [1–4].

4.2.1
Conformational Properties of Proteins

Generally, only 20 different amino acids are found in naturally occuring proteins. The physico-chemical properties of their side chains, such as size, shape, hydrophobicity, charge and hydrogen bonding, span a considerable range. They avoid, however, the extremes of high chemical reactivity and also, except for proline, strongly restricted degrees of freedom. The question most relevant in view of the 3D shape of proteins is, how the individual side chains interact with the backbone as well as with one another, and what roles they play within particular types of secondary and tertiary structures. The predominant influences of the sequential order on protein conformation are (aside from the linear connectivity and the steric volume) the hydrogen bonding capabilities and the chirality of all (except glycine) amino acid residues. All 19 chiral amino acids possess the L-configuration or according to the Cahn–

Ingold–Prelog scheme the S-configuration, with the exception of L-cysteine, which due to a change in ligand priority possesses the R-configuration.

An important convention needed for understanding much of the information available for a particular protein, is the designation of the individual atoms and structural elements of a protein. All atoms, angles and torsion angles that describe the 3D structure of a protein are named using letters in the Greek alphabet. The central carbon atom in amino acid residues is termed α, and the side chain atoms are commonly designated β, γ, δ, ε, and ζ in alphabetical order starting from the α carbon atom. The backbone of a protein consists of a repeated sequence of three atoms, belonging to one amino acid residue—the amino N, the C^{α} and the carbonyl C; these atoms are generally represented as N_i, C_i^{α}, and C_i' respectively, where i is the number of the residue, starting from the amino end of the chain. As an example, a portion of the backbone of a polypeptide chain is shown in Fig. 4.1. This illustrates the conventions used in describing protein conformation.

The main chain torsion angles in proteins are named ϕ (phi), ψ (psi) and ω (omega). Rotation about the N–C^{α} bond is described by the torsion angle ϕ, rotation about the C^{α}–C' bond by ψ, and rotation about the peptide bond C'–N by ω. Torsion angles of the side chains are designated by χ_j (chi$_1$, chi$_2$, etc.) where j is the number of the bond counting outward from the C^{α}-atom.

The peptide bond is usually planar because of its partial double bond character and nearly always has the *trans* configuration (ω = 180°) which is energetically more favorable than *cis* (ω = 0°). The *cis* configuration sometimes is found to occur with proline residues (about 10%). Small deviations from planarity of the *cis* or *trans* form with $\Delta\omega$ = –20° to 10° seem to be energetically acceptable.

The variations of ϕ and ψ are constrained geometrically due to steric clashes between neighboring but non-bonded atoms. The permitted values of ϕ and ψ were first analyzed and determined by Ramachandran et al. [5]. In their work computer models of small peptides were used to vary systematically ϕ and ψ with the purpose

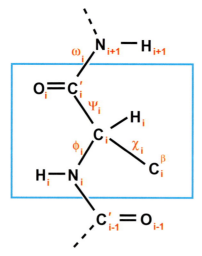

Figure 4.1 Designation of atoms and torsion angles in a protein.

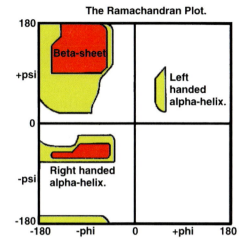

Figure 4.2 Ramachandran plot of a polyalanine.

of detecting stable conformations. Each conformation, represented by a particular ϕ, ψ combination, was examined for close contacts between atoms. In this rough model the atoms were treated as hard spheres with fixed geometries for the bonds. Only values of ϕ and ψ, for which no close contacts between atoms have been discovered, are permitted and usually are presented in a 2D map, the so-called Ramachandran plot. Since ϕ and ψ constitute a virtually complete description of the backbone conformation, the 2D Ramachandran plot is an important and easy-to-analyze test for the validity of 3D protein structure.

The Ramachandran plot of polyalanine is shown as an example in Fig. 4.2. The area outside the solid lines corresponds to conformations where atoms in the polypeptide chain are located in distances closer than the sum of their van der Waals radii. These regions are sterically disallowed for all amino acids, except glycine. Glycine, which lacks a side chain, is evenly distributed over the complete plain of the Ramachandran plot. The shaded regions correspond to conformations where no steric clashes are found, i.e. these are the allowed regions (or favored regions). The area directly outside the boundaries of this region includes conformations which are permitted if slight alterations of bond angles are accepted. Ramachandran plots for other amino acids appear comparable in the shape of the various regions.

Sub-regions of ϕ,ψ space are generally named after the secondary structure elements which result when the corresponding ϕ,ψ-angles occur repeatedly. The right-handed α-helix for example resides in the lower left near –60°, –40°, the broad region of extended β-sheets in the upper left around –120°, 140°, and the slightly unfavored left-handed α-helical region in the upper right near 60°, 40°. Conformational properties and other relevant parameters of these secondary structures are described in the following sections.

4.2.2
Types of Secondary Structural Elements

4.2.2.1 The α-Helix

The right-handed α-helix is the best known and most easily recognized secondary structural element in proteins [6, 7]. Approximately 32% to 38% of the residues in known globular proteins are involved in α-helices [8]. α-Helices are classified as repetitive secondary structure. That is, all C_α-atoms of α-helical amino acids are in identical relative positions. Thus, the ϕ,ψ torsion angle pairs are the same for each residue in the helix. The structure of an α-helix repeats itself every 5.4 Å along the helix axis; this means that the α-helices have a pitch of $p = 5.4$ Å. α-Helices have 3.6 amino acid residues per turn, i.e. a helix of 36 amino acids would form 10 turns.

The α-helical structure is mainly formed and stabilized by repeated hydrogen bonds between the carbonyl function of residue *n* and the NH of residue *n+4* (see Fig. 4.3). This results in a very regular and energetically favored state. α-Helices observed in protein structures are always right-handed. L-amino acids cannot form extended regions of left-handed α-helix because the C′-atoms would collide with the following turn. Only individual residues are found which possess the ϕ,ψ torsion

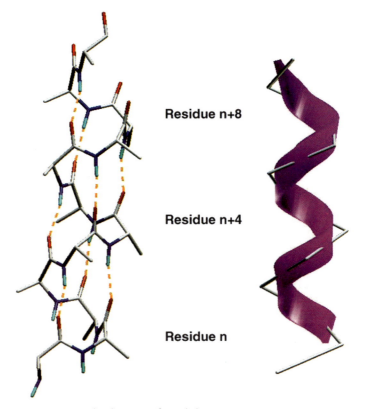

Residue n+8

Residue n+4

Residue n

Figure 4.3 General architecture of an α-helix.

angles of a left-handed α-helix. So, when speaking of an α-helix, usually the right-handed α-helix is meant.

The exact geometry of the α-helix is found to vary somewhat in natural proteins, depending on its environment. The ideal α-helix ($\phi = -57°$ and $\psi = -47°$) is only one version of a family of similar structures [6]. More usually, a slightly different α-helix geometry ($\phi = -62°$ and $\psi = -41°$) can be observed in proteins [7]. This conformation is more favorable than the ideal α-helix because it permits each carbonyl oxygen to make hydrogen bonds to both the NH of residue $n+4$ and the aqueous solvent (or other hydrogen bond donors).

The side chains of an α-helix are pointing outwards into the surrounding space. Several restrictions exist for side chain conformations, especially for side chains with branched C′-atoms (Val, Ile, Thr). Proline residues normally are incompatible with the α-helical structure because, due to the cyclic structure, the amide nitrogen lacks the hydrogen substituent necessary for hydrogen bonding. If single proline residues nevertheless appear in long α-helices (e.g. in some of the transmembrane α-helices of bacteriorhodopsin), this appearance yields a local distortion of the α-helical geometry.

Variations of the classical α-helix in which the protein backbone is either more tightly or more loosely coiled (with hydrogen bonds to residues $n+3$ and $n+5$), are named 3_{10}-helix and π-helix, respectively. In general, these helix types play only a minor role in the architecture of proteins. However, 3_{10}-helices frequently form the last turn of a classical α-helix.

4.2.2.2 The β-Sheet

Besides the α-helix, the second most regular and recognizable secondary structural motif is the β-sheet [9, 10]. Like the α-helix it is a periodic element. ,-Sheets are formed from β-strands which develop when a linear extended conformation of a polypeptide backbone ($\phi = -120°$, $\psi = 140°$) appears [9]. Since interactions between residues of the same strand are not possible, a β-strand is only stable as part of a more complex system, the β-sheet. As in α-helices all hydrogen bond donor or acceptor groups of the peptide backbone are engaged in the formation of hydrogen bonds, however, because these bonds appear not intra- but intermolecular, β-strands are energetically less favored. In contrast to α-helices—which consist of a singular stretch of directly bound residues—β-sheets possess a much more pronounced structure modulating effect because they are composed of several β-strands which can be distributed over a large part of the sequence.

Adjacent β-strands can be arranged in either parallel or antiparallel fashion. In parallel sheets the strands all run in the same direction (see Fig. 4.4(a)), whereas in antiparallel sheets they run in opposite directions (see Fig. 4.4(b)).

The side chains of β-strands are located nearly perpendicular to the plane of the hydrogen bonds (between the single strands). Along the strand they alternate from one side to the other. For antiparallel β-sheets typically one side is buried in the interior of the protein and the other side is exposed to the solvent. Therefore, the physico-chemical character of the amino acids tends to alternate from hydrophobic to hydrophilic. Parallel β-sheets, on the other hand, are usually buried on both sides,

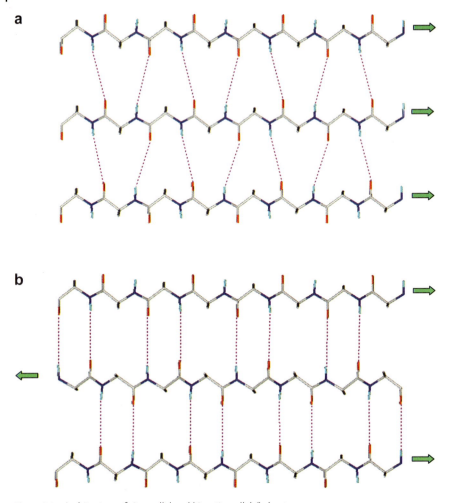

Figure 4.4 Architecture of a) parallel and b) antiparallel β-sheets.

so that the central residues tend to be hydrophobic, and hydrophilic amino acids are found abundantly towards the ends. For both types of β-sheet, edge strands can be much more hydrophilic than central strands.

β-Sheets are very common in globular proteins (20–28%) [8]. They can consist exclusively of parallel or antiparallel strands or are formed from a mixture of both. Purely parallel sheets are less frequent, while purely antiparallel sheets are very common. Antiparallel sheets often consist of as few as two or three strands, whereas parallel sheets always have at least four. Mixed sheets usually contain 3–15 strands.

4.2.2.3 Turns
Approximately one-third of all residues of globular proteins are involved in turn regions. The general function of turns is to reverse the direction of the polypeptide

chain. Often turns are located on the protein surface and therefore contain predominantly charged and polar amino acids.

Various different types of reverse turns have been observed in proteins. Their specific features depend for example on the type of secondary structural motifs which are linked by them. For a detailed description of all observed turn types the reader is referred to the literature [1–4, 11, 12].

Turns often connect antiparallel β-strands. In this case they are named ,-turns or hairpin bends [12]. Some 70% of hairpin turns are shorter than seven residues in length; most often they include only two residues. Larger loops have less well-defined conformations, which are often influenced by interactions with the rest of the protein. In all reverse turns the peptide groups are not paired by regular hydrogen bonds, but are accessible to the solvent. For this reason reverse turns often appear on the protein surface.

In general, the periodic secondary structural elements in proteins (α-helices and β-sheets) are rather short. The length of an α-helix is usually 10–15 residues (12–22 Å). A single β-strand is found to count 3–10 residues (7–30 Å). Most of the described ideal geometries of helices and sheets are only rarely observed in nature. Often, the geometries of secondary structures are more or less distorted. For example, solvent-exposed α-helices very often show a curved helix axis. Most β-sheets in folded proteins are rather twisted than planar with a twist of 0–30° between the single strands.

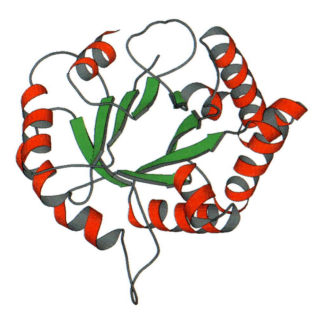

Figure 4.5   The 3D structure of triose phosphate isomerase presented in simplified form using MOLSCRIPT [13].

The common properties of proteins described here provide only some general rules of protein architecture. Each naturally occurring protein, on the other hand, is unique and attains its functional and structural character by means of specific non-covalent interactions. It is therefore necessary to compare each computer-generated structure with "real" 3D structures of proteins, and to include as much as possible information about protein structures in the process of protein modeling.

The exclusive presentation of secondary structural motifs of a complex protein in a schematic form is a very helpful tool for comprehending the overall structure. Usually in this kind of representation the side chains are omitted to yield a clearer picture of the whole protein, including the various secondary structural elements. Helices are often described by cylinders or coiled ribbons and extended strands of β-sheets by broad arrows indicating the amino-to-carboxy direction of the backbone. The 3D structure of triose phosphate isomerase is presented in such simplified form in Fig. 4.5.

4.2.3
Homologous Proteins

It has long been recognized that the evolutionary mechanism of gene duplication which is associated with mutations, leads to divergence and thereby to the foundation of families of related proteins with similar amino acid sequences and similar 3D structures. The proteins that have evolved evolutionarily from a common ancestor are said to be *homologous*. Two homologous sequences can be nearly identical, similar to varying degrees, or dissimilar because of extensive mutations. As a matter of fact the sequence similarity in homologous proteins is less preserved than the structural similarity. Or stated in a different way, 3D structures of homologous proteins have been remarkably conserved during their evolution because the common structure is crucial for the specific function of the proteins. The conservation of protein structure has been detected in many protein families. The 3D structures of α-chymotrypsin and trypsin, belonging to the family of trypsin-like serine proteases, can be cited as an example. They are remarkably similar, although they share only 44% identical amino acid residues. This topological similarity can easily be observed in Fig. 4.6. Other members of the family of serine proteases have changed more drastically during evolution. Bacterial serine proteases for example show only 20% sequence identity when compared with the mammalian enzymes like thrombin, trypsin or chymotrypsin. However, if the 3D structural similarity is considered the main features are still present.

The question which immediately comes to mind in this respect is how such large dissimilarities in the primary sequences are compatible with the observed structural similarity. The answer was found empirically and can be summarized as follows. The most pronounced dissimilarities generally appear in regions close to the protein surface, the so-called loop regions. In these regions even the physico-chemical properties of the side chains have often changed. Residues located in the interior of proteins, however, vary less frequently and less distinctly. This leads to the situation that generally a common core of residues comprising the center of the protein and

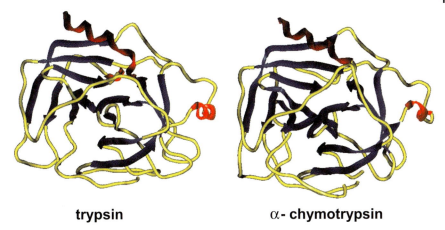

trypsin   **α- chymotrypsin**

Figure 4.6 3D structure of two homologous enzymes. Color code: red = α-helix; blue = β-strand; yellow = peptide backbone.

the main elements of secondary structure remain highly conserved within a family of homologous proteins.

Within homologous proteins the elements of secondary structure can move relative to each other, can change in length, or can even disappear completely. However, an α-helix is not usually replaced by a β-sheet, or vice versa. In general, neither the order nor the orientation (parallel or antiparallel) of β-strands has ever been recognized to differ between proteins of the same family.

In summary, the overall conformations of homologous proteins appear to have been highly conserved during evolution. This fact forms the basis for the development of the comparative protein modeling approach which will be described in the next section.

References

[1] Creighton, T. E. *Proteins: Structures and Molecular Properties,* 2nd Ed. W. H. Freeman and Company: New York 1993.

[2] Branden, C., and Tooze, J. *Introduction to Protein Structure.* Garland Publishing Inc.: New York 1991.

[3] Schulz, G. E., and Schirmer, R. H. *Principles of Protein Structure.* Springer Verlag: New York 1979.

[4] Fasman, G. D. *Prediction of Protein Structure and the Principles of Protein Conformation.* Plenum Press: New York 1989.

[5] Ramachandran, G. N., and Sasisekharan, V. *Adv. Prot. Chem.* **23**, 283–437 (1968).

[6] Pauling, L., Corey, R. B., and Branson, H. R. *Proc. Natl. Acad. Sci. U.S.A.* **37**, 205–211 (1951).

[7] Barlow, D. J., and Thornton, J. M. *J. Mol. Biol.* **201**, 601–619 (1988).

[8] Kabsch, W., and Sander, C. *Biopolymers* **22**, 2577–2637 (1983).

[9] Chou, K. C., Pottle, M., Nemethy, G., Veda, Y., and Scheraga, H. A. *J. Mol. Biol.* **162**, 89–112 (1981).

[10] Pauling, L., and Corey, R. B. *Proc. Natl. Acad. Sci. U.S.A.* **37**, 729–740 (1951).

[11] Rose, G. D., Gierasch, L. M., and Smith, J. A. *Adv. Prot. Chem* **37**, 1–109 (1985).

[12] Sibanda, B. L., and Thornton, J. M. *Nature* **316**, 170–174 (1985).

[13] Kraulis, P. J. *J. Appl. Crystallogr.* **24**, 946–950 (1991).

4.3

Comparative Protein Modeling

As we have already discussed in section 4.1 extensive information on primary and secondary structure of proteins are stored in various databases. Protein sequence determination is now routine work in molecular biology laboratories. As a result of the Human Genome project, the rate of publication of primary sequences has increased dramatically in the last few years. Sequences of more than 200 000 proteins are now available. The translation of sequences into 3D structure on the basis of X-ray crystallography or NMR investigations, however, takes much more time. Therefore, hitherto (by the end of 2002) the 3D structures of not more than 18 000 proteins are available in the Protein Data Bank. In certain circumstances it can take, depending on the kind of proteins studied, more than one year to perform a complete structure determination [1]. Therefore, many more protein sequences are known than complete 3D structures. Because of the technical problems related to experimental 3D structure elucidation, theoretical procedures for predicting protein 3D structure on the basis of the respective amino acid sequence are urgently needed. Since a general rule for the folding of a protein has not yet been developed, it is necessary to base structural predictions on the conformations of available homologous reference proteins [2–4] (see also section 4.2 for the underlying principles).

If one sequence is found homologous to another, for which the 3D structure is available, the comparative modeling approach (also called the homology modeling approach) is the method of choice for predicting the structure of the unknown protein. The underlying idea of comparative modeling is to make use of the collected body of knowledge about already resolved proteins. In a first step the sequence of a new protein is compared with all sequences of structurally known proteins stored in a database. Proteins in the database which are identified as homologous to the unknown are retrieved and used as templates for the structural prediction of the unknown protein. This approach was developed by several authors and is described in detail in the following paragraphs [5–8].

Successful comparative protein modeling, however, depends strongly upon how closely the structure that one is attempting to model fits the chosen template [9]. Because, at present, our understanding of protein folding patterns is still rather limited, the only criterion that can be applied for structure prediction is the examination of the extent of sequence homology between known and unknown protein. Although the conclusion of many studies in the past was that structural homology persists even if sequence homology is hardly detectable, for the purpose of knowledge-based modeling the reverse is important. The prediction of structural similarity between different proteins can only be based on the detection of homologies in their sequences. Thus, the comparison of sequences using alignment methods is a central technique in comparative modeling and will be described in detail in section 4.3.1.

The process of comparative modeling involves the following steps:

1. Determination of proteins which are related to the protein to be studied.
2. Identification of structurally conserved regions (SCRs) and structurally variable regions (SVRs).
3. Alignment of the sequence of the unknown protein with those of the reference protein(s) within the SCRs.
4. Construction of SCRs of the target protein using coordinates from the template structure(s).
5. Construction of SVRs.
6. Side chain modeling.
7. Structural refinement using energy minimization methods and molecular dynamics.

4.3.1
Procedures for Sequence Alignments

The first step in comparative modeling is the assignment of the unknown protein structure to a protein family. In many cases this information is already known because the sequence to be modeled belongs to a well-known protein family. However, this may not be true. Then it is necessary to compare the new sequence with thousands of sequences already stored in protein databases and to identify, if possible, homologous ones.

In the past, identifying new proteins through database searches has been difficult and time-consuming. Computer programs required several hours or made far-reaching compromises in sensitivity or selectivity of the search.

However, during the last decade heuristic methods were developed to tackle this problem [10–13]. These methods do not guarantee to always find the globally optimal solution, but in practice they rarely miss a particularly significant match. The two major methods are FASTA [12] and BLAST [13]. Both methods are implemented in many commercially available software packages (like HOMOLOGY [14], MODELLER [15], COMPOSER [16], WHAT IF [17], GCG [18]). In addition they are often integrated as search tools into databases and web sites with biological content.

The central technique used for amino acid sequence comparison is the so-called *sequence alignment*. In the framework of comparative modeling the sequence alignment procedure is of importance for several reasons. Firstly it is used to search databases in order to find related sequences and to identify which regions of the detected proteins are conserved, thus suggesting where the unknown protein may also be structurally conserved. This for example can be performed employing the above-mentioned FASTA, BLAST or PSI-BLAST programs. Secondly, sequence alignment is used for detection of correspondences between amino acids of the structurally known reference protein and those of the protein to be modeled. These correspondences are the basis for transferring the coordinates of the reference protein(s) to the model protein. For this task the more sensitive and selective alignment procedures described below are needed.

A very natural procedure for aligning sequences would be to simply write them in tabular form for visual inspection. Of course this would be not only unsystematic, it would be very time-consuming, especially if more than two sequences are to be compared. For that reason many programs have been developed which are able to perform alignments automatically [18–21]. Because the alignment of amino acid sequences is such a crucial step in homology modeling of proteins, many different methods and programs have been published and still are being developed. It is beyond the scope of this book to discuss all of them, but the reader is referred to the literature [12, 13, 18, 19].

One of the earliest attempts to clarify whether the structural similarity existing between proteins is due to homology or occurs by chance, was carried out by Needleman and Wunsch [20]. Variants of the algorithm used by these authors have been further developed independently by others and applied in many fields. These programs are more sensitive in detecting homology than the database search programs, but on the other hand are slower in finding an optimal alignment. However, the great advantage of the Needleman and Wunsch algorithm is that final detection of the best alignment for two sequences is guaranteed. As a consequence computer programs based on this method (ALIGN, BESTFIT and GAP which are included in the GCG program package [18]) have been widely used for biological sequence comparison. Whereas the original Needleman and Wunsch algorithm is only able to align two sequences, many up-to-date programs handle the alignment of more than two sequences. These so-called multiple alignment methods are significantly more difficult than the pairwise alignment techniques. This is because the number of possible alignments increases exponentially with the number of the sequences to be compared. Several programs have been derived to provide approximate solution of this problem (for example CLUSTALW [18] or MAXHOM [21])

In contrast to the above-described procedures—which search for the global optimal similarity of sequences—other approaches seek to identify the best local similarities between two sequences. These so-called *optimal local alignment* methods are likewise based on a modified Needleman and Wunsch algorithm and represent an important tool for comparing sequences. This is especially true for the location of highly homologous regions dispersed over long sequences [22–24]. The basic idea of these methods is to consider only relatively conserved subsequences of homologous proteins; dissimilar regions do not contribute to the measure (Fig. 4.7).

In the course of comparising of two sequences the alignment procedures, at least in effect, seek to duplicate the evolutionary process involved in converting one sequence into another. For this operation a kind of scoring scheme is required that dictates the weight for aligning a particular type of amino acid with another. This type of scoring scheme is provided by the so-called *homology matrices*, which make use of the most probable amino acid substitutions according to physical, chemical or statistical properties. High numerical values in the matrix imply that a substitution is probable, whereas low values indicate that a substitution is unlikely to occur. From the various kind of matrices which are in use [25–29] the most often applied are:

```
ENTCL : PVSEKQLAEVVANTITPLMKAQSVPGMAVAVIY--QGKPHYTFGKADIAANKPVTPQILFELGSISKTFTGVLGGDAIA-
CITFR : AKTEQQIACIVNRTITPLMQEPAIPGMAVAIIY-EGKPYYTFWGKADIANNHPVTQQTLFELGSVSKTFNGVLGGDRIA-
MEN1  : QTADVQQKLAELERQSG-GRLGVALINTADNSQILYR--------ADERFAMCSTSKVMAAVAAVLKKSE-
STAU  : KELNDLEKKYN-AHIGVYALDTKSGKEVKFN---------SDKRFAYASTSKAINSAILLEQVP-
BALI  : DDFAKLEEQFD-AKLGIFALDIGTNRTVAYR-----------PDERFAFASTIKALTVGVLLQQKS-


ENTCL : -RGEISLDDAVTRYWPQLTGKQWQ--------GIRMLDLATYTAGGLPLQVPDEVTDNASLLRFYQNWQPQWKPGTTRLYANASIGLFGALAVKPSGMPYE
CITFR : -RGEIKLSDPVTKYWPELTGKQWR--------GISLLHLATYTAGGLPLQIPGDVTDKAELLRFYQNWQPQWTPGAKRLYANSSIGLFGALAVKSSGMSYE
MEN1  : -SEPNLLNQRVEIKKSDLVNPIAEKHVDGTMSLAELSAAALQ-------------YSDNVAMNKLISHVGGP--ASVT
STAU  : --YNKLNKKVHINKDDIVAYSPILEKVVGKDITLKALIEASMT---------------YSDNTANNKIIKEIGGI--KKVK
BALI  : ---IEDLNQRITYTRDDLVNPITEKHVDTGMTLKELADASLR----------------YSDNAAQNLILKQIGGP--ESLK


ENTCL : QAMTRVLKPLKLDHTWINVPKAEEAHYAWGYRDGKAVRVSPGMLDAQAYGVKTNVQDMANWVMANMAPENVADASLKGGIALAQSRYWRIGSMYQGLGW
CITFR : EAMTRRVLQPLKLAHTWITVPQSEQKNYAWGYLEGKPVHVSPGQLDAEAYGVKSSVIDMARWVQANMDASHVQEKTLQQGIELAQSRYWRIGDMYQGLGW
MEN1  : AFARQLG-----DETFRLDREPTLNTAIPGDPRD---------------TTSPRAMAQLRNLTLGKALG----DSQRAQLVTWMKGNTTGAASIQA
STAU  : QRLKELG-----DKVTNPVRYEIELNYYSPKSKKD-------------TSTPAAFGKTLNKLIANGKLS----KENKKFLLDLMLNNKSGDTLIKD
BALI  : KELRKIG-----DEVTNPEREFEPELNEVNPGETQD------------TSTARALVTSLRAFALEDKLP-----SEKRELLIDWMKRNTTGDALIRA


ENTCL : EMLNWPVEANTVVEGSDSKVALAPLPVAEVNPPAPPVKASWVHKTGSTG--GFGSYVAFIPEK----QIGIVMLANTSY-----PNPARVEAAYHILEAL
CITFR : EMLNWPLKADSINGSDSKVALAALPAVEVNPPAPAVKASWVHKTGSTG--GFGSYVAFVPEK---NLGIVMLANKSY-----PNPARVEAAWRILEKL
MEN1  : GLPAS-------------------WVGDKTGSGD-YGTTNDIAVIWPKD-RAPLILVTYFTQPQPKAESRRDVLASAAKIVTNGL
STAU  : GVPKD--------------------YKVADKSGQAITYASRNDVAFVYPKGQSEPIVLVIFTNKDNKSDKPNDKLISETAKSVMKEF
BALI  : GVPDG--------------------WEVADKTGAAS-YGTRNDIAIIWPPK-GDPVVLAVLSSRDKKDAKYDDKLIAEATKVVMKAL
```

Figure 4.7 Multiple sequence alignment of cephalosporinases from *Enterobacter cloacae* (ENTCL) and *Citrobacter freundii* (CITFR), with penicillinases from *Escherichia coli* (MEN1), *Bacillus licheniformis* (BALI) and *Staphylococcus aureus* (STAU). Red letters indicate the determined SRCs.

1. *Identity matrix*: this is the simplest matrix that gives a score of 1 to identical pairs and 0 to all others.
2. *Codon substitution matrix*: the scoring values for this matrix are derived from the DNA base triplets coding for the amino acid pairs. For each pair, all of the possible nucleotide triplets are examined and the number of point mutations required to change one amino acid into the other are evaluated. Identical amino acids get a score of 9, one required mutation gives a score of 3, and two mutations yield a score of 1.
3. *Mutation matrix* (also known as the Dayhoff or PAM250 matrix [25]): this matrix was obtained by counting the number of substitutions from one particular amino acid by others observed in related proteins across different species. Large scores are given to identities and substitutions which are found frequently, and low scores are assigned to mutations that are not observed. Due to this procedure larger scores are used for certain non-identical pairs than for some identical ones. The Dayhoff matrix (Fig. 4.8), is the most widely used scoring scheme. It is often applied for finding an initial alignment for two unknown sequences. An advanced form of the Dayhoff matrix was suggested by Gribskov et al. [26]. The Gribskov matrix assigns the highest score always to identical amino acid pairs.
4. *Physical property matrices*: the scores of corresponding matrices are based on similarity indices for certain physical properties of amino acids, such as hydrophobicity, polarizability or helical tendency [28].

Differences in sequence lengths or variations in the locations of conserved regions complicate the alignment procedure. If one or both of the mentioned pro-

	Ala	Arg	Asn	Asp	Cys	Gln	Glu	Gly	His	Ile	.
Ala	2	-2	0	0	-2	0	0	1	-1	-1	
Arg	-2	6	0	-1	-4	1	-1	-3	2	-2	
Asn	0	0	2	2	-4	1	1	0	2	-2	
Asp	0	-1	2	4	-5	2	3	1	1	-2	
Cys	-2	-4	-4	-5	12	-5	-5	-3	-3	-2	
Gln	0	1	1	2	-5	4	2	-1	3	-2	
Glu	0	-1	1	3	-5	2	4	0	1	-2	
Gly	1	-3	0	1	-3	-1	0	5	-2	-3	
His	-1	2	2	1	-3	3	1	-2	6	-2	
Ile	-1	-2	-2	-2	-2	-2	-2	-3	-2	5	
.											

Figure 4.8   Dayhoff evolutionary mutation matrix.

blems are found, gaps are introduced into the sequence to allow the simultaneous alignment of all conserved regions. To limit the total number of inserted gaps (a large number would render the alignment increasingly unrealistic), an additional factor is implemented into the alignment algorithms, the so-named *gap penalty function*. The overall balance between the number of aligned amino acids and the smallest number of required gaps leads to an optimal alignment.

The combination of an alignment algorithm, a scoring matrix, and a gap-weighting function constitutes a system which can optimally align two or more sequences. The quality of a particular alignment is described by the *alignment score*. It is important to know that a derived alignment for related sequences is optimal only for the chosen parameters; changing the values can lead to a different alignment and a different score. Thus, it should be borne in mind that automatic sequence alignment methods are far from being perfect. The resulting alignment should always be verified for reasonableness. All known information on all levels on protein organization (primary, secondary and tertiary structure) have to be incorporated in the examination. Only when the derived alignment agrees with all known structural data can it be used as a basis for the generation of a protein model.

Another fundamental problem of all sequence alignments is found in the fact that recognizable sequence homology is lost more rapidly during evolution than the underlying structural similarity. Thus, it is difficult to give simple rules for the degree of similarity necessary to demonstrate unambiguously that two protein sequences are homologous. This depends strongly on the lengths of the sequences and their amino acid compositions. During the past decade several investigations have been performed to quantify the relation between sequence and structural homology [30–32].

Doolittle has defined some rules of thumb which can ease the decision [30]. If the sequences are longer than 100 residues and are found to be more than 25% identical (with appropriate gaps) then they are very likely related. If the identity is in the range of 15–25%, then the sequences may still be related. If the sequences are less than 15% identical, they are probably not related.

In order to be able to take a decision in the undecided range between 15–25% homology it must be proven that the alignment is statistically meaningful. One way to evaluate this point is by comparing the actual alignment score, which reflects the amount of homology between two sequences, with the average alignment score of randomly permuted sequences (which were generated by randomly exchanging the amino acid residues in the original sequences). This procedure preserves the exact length and amino acid composition of the proteins, and the statistical variation of the random comparison provides a measure of the significance of the observed similarity. A number of n randomizations for both sequence 1 and 2 will be generated. Each derivative of sequence 1 is then aligned against each derivative of sequence 2, resulting in a total of n^2 alignments. Both the mean and the standard deviations of the alignments are normally reported and can be compared with the original score. As an approximate guide; if the alignment score is more than six times the standard deviations above that for the random alignment, most of the residues in secondary structures will be correctly aligned [31].

Chothia and Lesk have performed an investigation on homologous proteins in order to quantify the relation between sequence homology and 3D similarity in core regions of entirely globular proteins [32]. They have found, that the success to be expected in modeling the structure of a protein from its sequence (using the 3D structure of a homologous protein as template) depends to a high degree upon the extent of sequence identity. They concluded that a protein structure provides a close general model for other proteins if the sequence identity is above 50%. If the sequence homology drops to 20%, large structural differences can occur (see Fig. 4.13). However, they found that the active site of distantly related proteins can have very similar geometries. Thus, in cases where the sequence identities are low, the structure of the active site in a protein may provide a reliable model for those in related proteins.

4.3.2
Determination and Generation of Structurally Conserved Regions (SCRs)

Building a protein model using the homology approach is based on the fact that there are regions in all proteins belonging to the same protein family that are nearly identical in their 3D structures. These regions tend to be located at the inner cores of proteins where differences in peptide chain topology would have significant effects on the tertiary structure of the protein [33]. Accordingly, it has been observed that the secondary structural units of strongly related proteins, above all α-helices and β-strands, occupy the same relative orientations throughout the whole protein family. As a natural follow-up these regions lend themselves to being used as the basic framework for the assignment of atomic coordinates for one of the other proteins belonging to the same family. These segments are called *structurally conserved regions* (SCRs).

The accurate assignment of SCRs within a family of homologous proteins is affected by several factors. The way to proceed depends on the number of available crystal structures of homologous proteins. It is fortunate when more than one crystal structure at atomic resolution is available. In this situation one can examine all structures in order to discover where the proteins are conserved structurally, even with regard to the 3D structure. To recognize the conserved parts of the proteins, they must be superimposed relative to each other. This is normally done using least-squares fitting methods. The main problem in this context is the selection of the corresponding fitting atoms; this means that it is not known a priori which part of the protein should be aligned to receive the best 3D overlap. In a first approximation the structures can be superimposed by least-square fitting of the C^α-atoms [3]. The initial superposition then can be optimized using only matching points located in secondary structural elements that are found to be conserved. Several approaches have been developed which try to solve the fitting problem automatically [34–40].

Matthews and Rossmann [40] have suggested a method which uses the least-squares fitting procedure. In a first step, two protein structures—which have to be aligned—are least-squares fitted using an initial set of equivalent residues. The equivalences are then updated according to both the distances between potentially equivalent residues and local directions of the main chain. The superposition and updating is repeated until no increase can be obtained in the number of equivalences.

In general, the resulting superimposed 3D structures show that large parts of the two proteins are very similar in structure and hence appear to be the structurally conserved regions, while other sections differ considerably. It should be noted that the applied algorithms do not take into account explicitly the secondary structure. Since—according to the definition SCRs must be terminated at the end of a secondary structural unit, so that, for example, each single strand of a β-sheet comprises a separate SCR—secondary structural elements of the proteins must be assigned before SCRs are determined. Information about the secondary structure of any known protein can be derived in the easiest way from crystal data files (for example from the PDB files) which include the secondary structural elements detected by crystallographers. Because the assignment of secondary structures in crystal structure files is often subjective and sometimes incomplete, it is more convenient to use objective methods which are able to assign correctly the secondary structural elements. Programs like DSSP [41] or STRIDE [42] detect secondary structural elements on the basis of geometrical features, i.e. the hydrogen bonding pattern or the main chain dihedral angle. Using these programs—which are accessible via the EBML web site—one can rapidly assign secondary structures to all proteins if atomic coordinates exist.

The situation is more complicated when only a single homologous protein is known that can be used as reference structure for the target sequence, because with only one known template protein a basis for a structural comparison does not exist. Under these circumstances one has to detect the SCRs manually using both, sequence and structural information of the proteins. As was described before, conserved regions are frequently detected in stable secondary structure elements. Therefore, it is reasonable to study carefully as many of those elements as possible in the reference protein with the aim of discovering potential clues for the existence of SCRs. Residues in the hydrophobic core tend to be more conserved with regard to sequence and 3D structure than residues at the protein surface. Amino acids involved in salt bridges, hydrogen bonds and disulfide bridges are most likely to be conserved within a protein family. The same holds true for amino acids located in the active site. Information derived from multiple sequence alignments can also be used beneficially to locate the SCRs more accurately.

It was found in many investigations on homologous proteins, that SCRs show strong sequence homology, while the variable regions show little or no sequence homology and are the sites of addition and deletion of residues. For that reason the determined SCRs should have identical or closely homologous sequences. Due to the structural homology of these regions no gaps are allowed in conserved areas.

In cases where the SCRs of the reference proteins already are known one has only to locate the regions of the model protein that correspond to these SCRs. This is accomplished by aligning the target sequence with the sequences of the SCRs in the homologs. The alignment procedure which must be applied for this purpose differs slightly from that already described. Because, by definition, SCRs cannot contain insertions or deletions, an algorithm is needed which disallows the introduction of gaps within SCRs. Unfortunately the standard Needleman and Wunsch method does not have the measure for treating SCRs in a special manner. It places a gap at any location if this results in an optimized amino acid matching. For this reason procedures have been developed [3, 22, 43] which can handle each SCR independently. Corresponding programs generate alignments without gaps appearing within any conserved region. When the correspondence between the reference and the target sequences has been established the coordinates for the SCRs can be assigned. The coordinates of the reference proteins are used as basis for this assignment. In segments with identical side chains detected in reference and target proteins, all coordinates of the amino acids are transferred. In diverse regions only the backbone coordinates are transferred. The corresponding side chains then will be added after complete backbone (SCRs and SVRs) generation (see section 4.3.4).

4.3.3
Construction of Structurally Variable Regions (SVRs)

Since significant differences in protein structures occur preferably in loop regions, the construction of these *structurally variable regions* (SVRs) is a more challenging task. Insertions and deletions due to differences in the number of amino acids additionally complicate the modeling procedure. A variety of methods for generating loops have been developed and described comprehensively in the literature [5–7, 44–46]. A good guide for modeling the missing region can be the structure of a segment of equivalent length in a homologous protein. Extensive investigations of variable regions in homologous proteins have shown that in cases where particular loops possess the same length and amino acid character, their conformation will be the same. The coordinates then can be transferred directly to the model protein in the same way as described for the SCRs. If no comparable loop exists in the protein family, two other strategies can be applied for modeling the SVRs. The coordinates for the SVRs can be either retrieved from peptide segments which are found in other proteins and that fit properly into the model's spatial environment [5–7], or by generating a loop segment de novo [44–46]. The former approach, which is known as *loop search method*, looks for peptide segments in proteins which meet the specified geometrical criterion. Usually the loop search programs are scanning the Protein Data Bank for possible peptide segments. The specified geometry input for the database search is given by distances and coordinates, including the residues of the regions embracing the loop segment in the model. The output of a respective search is a collection of loops satisfying the specific geometrical constraints. Usually the 10 to 20 best loop fragments are retained for further examination. The loops are ranked according to goodness of fit to the desired structure. However, additional criteria not used explicitly during the loop search, can provide a guide to ascertain the preference of one loop candidate over another. The retrieved fragments can be analyzed on the basis of quality of fit to the residues confining the loop region, by determining sequence homology between the original loop sequence and the sequence of the retrieved fragment, or via evaluation of steric interactions and energy criteria.

The loop search method offers the advantage that all loops found are guaranteed to possess reasonable geometry and resemble known protein conformations. It is not certain that the chosen segment fits properly into the existing framework of the model, so severe sterical overlaps may be detected. If this happens, the *de novo generation technique* is an alternative method.

Using this approach a peptide backbone chain is built between two conserved segments using randomly generated numerical values for all the backbone dihedral angles. Several algorithms have been developed to optimize the search strategy and to reduce computing time. Due to the complexity of this type of search method the approach can only be used for loops smaller than seven residues.

All loops generated by database or random search methods are usually far from optimal geometry. For that reason all loop regions (including confining residues) must subsequently be refined by energy minimization techniques in order to remove steric hindrance and to relax the loop conformations (see section 4.4.3).

4.3.4
Side Chain Modeling

When the peptide backbone has been constructed the next step is to add the side chains. The prediction of the numerous side chain conformations is by far a more complex problem than the prediction of the backbone conformation of a homologous protein. Many of the side chains possess one or more degrees of freedom and therefore can adopt a variety of energetically allowed conformations.

Several strategies have been developed in the past to find a solution for this multiple minima problem [47–54]. It has been generally assumed that identical residues in homologous proteins adopt similar conformations. Also, when the substituted side chain belongs to an amino acid pair that shows high homology (indicated by a high score in the Dayhoff matrix, for example Ile and Val, or Gln and Glu), it is assumed that the side chains adopt the same orientation in the protein [47]. The situation will become more complicated if the amino acids to be substituted are not related. When the side chain to be considered is longer than its counterpart in the homologous protein or is structurally dissimilar, the side chains must be positioned at random but in a conformation that avoids unfavorable contacts with other side chains [48]. An alternative way to obtain a suitable side chain conformation is to select the calculated minimum conformations of the appropriate dipeptide potential energy surface [49].

A more reliable procedure was developed by examining the relationship between the side chain positions in homologous structures of globular proteins. It has been found that the side chains adopt usually only a small number of the many possible conformations [50, 51]. Side chains with for example two χ angles have been observed to exist in four to six common conformations. All observed rotamers are combinations of the familiar gauche and anti forms. On the basis of such statistical evaluations rotamer libraries have been developed [50, 53]. An often applied side chain library is the one created by Ponder and Richards [50] which contains 67 rotamers for 17 amino acids. Several homology modeling programs make use of this library for generating the side chains of homologous proteins. Selecting the most probable conformation out of a rotamer library for side chain modeling might be problematical because this procedure disregards the information that is available from the equivalent side chain of the reference structure. Apart from that, the correct conformation of a side chain depends essentially on the local environment met by the amino acid in the real protein. This has been shown by several authors who have investigated well-resolved protein structures [54, 55]. In the interior of a protein, hydrophobic interactions are predominant and result in tight packing of amino acid residues. Factors such as the secondary structure and tertiary contacts with other residues can influence the side chain conformation. For that reason, methods have been developed which take into account information about the local environment and other constraints which may determine the positions of side chains. Sutcliffe et al., for example, have developed rules for mutual substitution of all 20 naturally occurring side chains in α-helical, β-sheet and loop regions—a total of $20 \times 20 \times 3 = 1200$ rules [54]. In order to determine which atom positions are preserved when substituting one amino acid for another at a topologically equivalent position, the study was performed on several sets of homologous proteins. All residues corresponding to a particular topologically equivalent position were aligned on their backbone atoms and inspected to determine which atoms are correlated in spatial position.

A further refinement of the Ponder and Richards approach has been developed by Dunbrack et al. [56]. The so called SCWRL program takes into account that side chain conformations depend upon the conformation of the main chain. However, all available side-chain prediction methods invariably keep the backbone fixed.

As we have discussed, various methods for the modeling of side chains do exist. All of them can greatly assist the modeler by providing appropriate side chain conformations. On the other hand, in several situations one has to refine side chain positions manually. Modifications must be applied, for example, when amino acids are involved in specific interactions like ion-pair formations, disulfide bridges, buried charge interactions or internal hydrogen bonds. Variations also occur when the residues are located on the protein surface and are fully accessible. Such exceptions must be treated on a case-by-case basis.

Once the final model has been built, a refinement of the structure is usually desirable. Regions where SCRs and SVRs are connected usually suffer from high steric strain and must be minimized. Several side chains may also adopt positions which result in bad van der Waals contacts. A stepwise approach for the structure refinement is likely to produce the best result. Overall simultaneous optimization of all side chains possibly would destroy important internal hydrogen bonds and may cause a conformational change within conserved regions. In order to remove steric overlaps, conformational searches are applied for residues which show bad van der Waals contacts. Energy minimizing and/or molecular dynamics of the model are useful routes to explore the local region of conformational space and may produce a more refined structure. The details about energy minimization and molecular dynamics used for structure refinement will be described comprehensively in section 4.4.3.

4.3.5
Distance Geometry Approach

While several reference structures are often used in the traditional homology modeling process, only one set of coordinates can be used for the construction of a particular structurally conserved region (see section 4.3.2). The distance geometry approach in comparative modeling [38, 57, 58] offers the possibility to examine all the reference proteins simultaneously to impose structural constraints that in turn can be used to generate conformations consistent with the data set. The first step using this procedure is the same as in the traditional homology modeling approach. The SCRs are identified and the sequence of the target protein is aligned with the sequences of the known proteins. The distance geometry method applies rules by which a multiple sequence alignment can be translated into distance and chirality constraints, which are then used as input for the calculation. By this means one obtains an ensemble of conformations for the unknown structure, where each member of the ensemble contains regions (which were constrained during the calculation) showing similar conformations and regions (which were free during the calculation) with varying arrangements. The structures of the ensemble then are energy-minimized in order to eliminate structural irregularities that sometimes appear during distance geometry calculations. The differences among the derived conformations provide an indication of the reliability of the structure prediction. A detailed description of this technique is given in a study reporting the application of the method to predict the structure of flavodoxin from *E. coli* [59].

4.3.6
Secondary Structure Prediction

The best method for the generation of a structure proposal for a protein with unknown 3D structure is to base it on a homologous protein whose 3D structure *is* available, i.e. by means of the knowledge-based approach as described earlier. However, in cases where a homologous protein does not exist, several other methods have been developed that have concentrated on the prediction of secondary structure. The underlying idea evolves from the fact that 90% of the residues in most proteins are engaged either in α-helices, β-strands or reverse turns. As a consequence it seems possible—if the secondary stuctural elements are predicted accurately—to combine the predicted segments in an effort to generate the complete protein structure. Obviously the reliability of this approach is much lower than homology modeling, thus, it should be applied with extreme caution. However, the prediction of secondary structure from the amino acid sequence has been widely practiced (for reviews see [60–66]).

Basically, three different types of methods can be employed for this task: statistical, stereochemical and homology/neural network-based methods. All different prediction methods rely, more or less, on information derived from known 3D structures stored in the Protein Data Bank. The correct assignment of secondary structural regions in the crystal structure (see section 4.3.2) is therefore necessary for a reliable validation of all prediction methods.

Statistically based methods were among the first that have been developed. The underlying idea takes advantage of the observation that many of the 20 amino acids show statistically significant preferences for particular secondary structures. Ala, Arg, Gln, Glu, Met, Leu and Lys for example are preferentially found in α-helices, whereas Cys, Ile, Phe, Thr, Trp, Tyr and Val occur more frequently in β-sheets. The most simple and most commonly used statistical method for secondary structure prediction is the one proposed by Chou and Fasman [61]. The prediction is done by calculating the probability of an amino acid to belong to a particular type of secondary structure, such as α-helix, β-sheet or turn, based simply on its frequency of occurrence as part of the respective secondary structure elements as found in the Protein Data Bank. Another commonly used statistical-based method is that of Garnier, Osguthorpe and Robson (GOR) [62]. The success of this type of algorithm is difficult to verify because some of them merely produce tendencies towards a particular secondary structure rather than an absolute prediction. Therefore, the methods are open to divergent interpretations, with the result that different authors obtain different results. The scope and limitations of the statistical methods have been demonstrated by Kabsch and Sander [69] in an analysis of three commonly used prediction methods showing that all methods are below 56% accurate in predicting helix, sheet and loop.

Another type of secondary structure prediction method is based on the interpretation of the hydrophobic, hydrophilic and electrostatic properties of side chains in terms of the formulation of rules for the folding of proteins [64–66]. The method of Lim, for example, takes into account the interactions between side chains separated by up to three residues in the sequence [64] in view of their packing behavior in either the α-helical or β-sheet conformations. A sequence with alternating hydrophobic and hydrophilic side chains, for example, is likely to be found in a β-sheet strand, with hydrophilic residues exposed to the solvent and hydrophobic residues buried in the interior of the protein. Correspondingly, the stereochemical-based methods have been applied successfully for the prediction of amphiphilic helices [65] or membrane-spanning segments [66].

Rost and Sander have reported an algorithm which uses evolutionary information contained in multiple sequence alignments as input to neural networks [67, 68]. Neural networks potentially have a methodological advantage compared with other prediction methods because they can be trained. This means that rules determining the behavior of the studied systems are not needed in advance, but are formed by the network itself on the basis of known facts.

The neural network method (called PHD) showed more than 70% accuracy in the prediction of three classes of secondary structure (helix, sheet, loop) on the basis of only one known homologous sequence [68, 69]. Other neural network-based prediction methods have also been reported to reach up to 80% accuracy [70,71]. The neural network methods PHD [69] and PSIPRED [70] are currently the methods of choice for the prediction of unknown sequences and are integrated into several bioinformatics-related web sites. An evaluation of secondary prediction methods can be found on the EVA web server (http://cubic.bioc.columbia.edu/eva)

Information derived from secondary structure prediction of homologous proteins is often used in addition to the results received in a primary sequence alignment in

```
  1              50
SEQUENCE      MMRKSLCCALLLGISCSALATPVSEKQLAEVVANTITPLMKAQSVPGMAV

CHOU                EEEEEEEEE    HHHHHHHHHHHHH   HHHHHHHHHHTTEEE
GOR           H   HH      E                HHHHHHHH HH   HH       EEE
ALB             HHHHHHHHHHHTTTTT  TTT    HHHHHHHHHHHHHHH    TTEEE
JAMSEK        HHHHHHHH        TTT        HHHHHHHHTTT              E
PHD           TTTT HHHHHHHHHHHHH    T   HHHHHHHHHHHHHHHH    TTTTT EE
DSSP                              HHHHHHHHHHHHHHHHHHHH    EEEEEEE


 51             100
SEQUENCE      AVIYQGKPHYYTFGKADIAANKPVTPQTLFELGSISKTFTGVLGGDAIAR

CHOU          EEEEE     EEEEE HHHHH   EEEEEEE        EEEEEETTTHHHH
GOR           EEEE      EEE HHHHH      EHEEHE H  EEEEEE      EH
ALB           EEEE TTTTEEEE          TT TT      HHHHHHHHHHHHTTHHHH
JAMSEK        EEEE TTT EEEETTT     TTT   TTT         TTTEEEE  HHHHHH
PHD           EEEE TT EEEE     TTTTTTTTT                       H
DSSP          ETTEEEEEEEEEEETTTT   EE TTTTEEEE      HHHHHHHHHHH


101             150
SEQUENCE      GEISLDDAVTRYWPQLTGKQWQGIRMLDLATYTAGGLPLQVPDEVTDNAS

CHOU          HHHHHHHEEEEEE    TT     HHHHHHHH  TT EEEEETT  TTTTT
GOR            HE  HHHEE E HH      HHHEEHHHH  H            HHH HH
ALB             THHHHHHHHH                    TTTEEEEETTT   TTHH
JAMSEK        HH HHHHHHHHH TTT  TTTHHHHHHHHHEEE         TTT   TTHH
PHD               TTT     TT H       T      HHHHH TTTTT TT   H HHH
DSSP              TTTT        TTTTTT     TTT HHHHHH  TTTTTTTTTTTT  HHH


151                          200
SEQUENCE      LLRFYQNWQPQWKPGTTRLYANASIGLFGALAVKPSGMPYEQAMTTRVLK

CHOU          EEEEEEEETT   TTEEEEEEEEEEEEEHHHHHHTTT    HHHHHHHHHH
GOR           HHEEHH              EEEHH  HHHHHH         HHHHHHHHH
ALB           HHHHH    TTT TTT         HHHHHHHH   TTT HHHHHHHHHHH
JAMSEK        HHHHHHTTTTTTTTT    EEETTT EEEEE    TTT        HHHHHH
PHD           HHHHHHH TTTTTTTT EE   TT    HHHHHH    TTT HHHHHHHHHHHH
DSSP          HHHHHHH       TTTTEE   HHHHHHHHHHH      HHHHHHHHH


201             250
SEQUENCE      PLKLDHTWINVPKAEEAHYAWGYRDGKAVRVSPGMLDAQAYGVKTNVQDM

CHOU          HHHHHHEEEEEHHHHHHH      TTTHHHHHHTTHHHHHH    EEEEEEHH
GOR            H H HHHHH  HHHHHHHHH H      EEE     HHHHHHH   E HHHH
ALB                EEEEETTT              TTT EEE HHHHHHHHHHHHHHHHHHH
JAMSEK        HHHTTTEEEEE          TTT    EEE              EEEEE
PHD           H TTTT    TTHHHHHHHHH  TTTT EE TTT T    TTT  HHHHH
DSSP            TTTEETTT          EETTTTEE   TTTHHHHHHEEEEHHHH
```

Figure 4.9 Comparison of secondary structure predictions using different methods for a crystallographically resolved cephalosporinase from *Enterobacter cloacae*. Structure elements shown in red agree with the structures observed in the crystal (H = α-helix, E = β-strand, T = turn)

```
      251           300
SEQUENCE      ANWVMANMMAPENVADASLKQGIALAQSRYWRIGSMYQGLGWEMLNWPVEA

CHOU          HHHHHHHHHHHHHHHHHHHH EEEEEEEEEEE   EEEEE HHHHHHHHHH
GOR           HHHHHH H   H HHHHHHHHHHHHHH  EEEEEEEE      HHHE
ALB           HHHHHHHH HHHHHHHHHHHHHHHHHHHHH       HHHHHHHHHHHH
JAMSEK          EEE       HHHHHHHHHHHHHEEEETTTEEEEEETTT              T
PHD           HHHHHH  T   T  HHHHHHHHHHH                 T       TTTTT
DSSP          HHHHHHHH        HHHHHHHHHHH EEEEETTEEETTTTEEEETTT H


      301           350
SEQUENCE      NTVVEGSDSKVALAPLPVAEVNPPAPPVKASWVHKTGSTGGFGSYVAFIP

CHOU          HHHHH TTHHHHHHHHHHHHHH TT    HHHHHH TTTTTTTEEEEEEEE
GOR            EEE    HHH     H        H  EEEEE      E EEEEE
ALB           EEEETTTTEEEEE EEEEEE TT                    EEEEET
JAMSEK        TT      TTT        TTT                      EEEE
PHD              TTT       TTTTT    TTTTT      E  TT  T   EEEEE
DSSP          HHHHHHH HHHHH   EE EEEEEE   TTTEEEEEEEEEETTEEEEEEEE


      351           381
SEQUENCE      EKQIGIVMLANTSYPNPARVEAAYHILEALQ

CHOU             EEEEEEEEEEETTTTTTTHHHHHHHHHHHHHH
GOR           HHHHHEEEE           HHHHHHHHHHHHH
ALB           TTEEEEEEE           TT HHHHHHHHHHH
JAMSEK          EEEEE   TTT
PHD             EEEEE   TTTTTHHHHHHHHHHHHHH T
DSSP            EEEEEEE        HHHHHHHHHHHHHH
```

Figure 4.9 continued.

order to improve the location of the SCRs in a class of homologous proteins. Even when only the structure of one homologous protein is known (which can be used as template for the comparative protein modeling approach), but several homologous sequences, it is helpful to include the predicted secondary structural elements for the homologous sequences to assign the SCRs. All available prediction methods should be applied in order to find the most probable assignment for the secondary structural elements. Of course different methods do not yield exactly the same result. This is shown in Fig. 4.9 using five methods (CHOU, GOR, ALB, JAMSEK, PHD) for the prediction of the known secondary structure of a cephalosporinase from *Enterobacter cloacae*. The prediction is also compared with the result of the DSSP program, which assigns the secondary structure on the basis of the known atomic coordinates.

Most of the prediction methods described are implemented in commercially available protein modeling programs or are integrated into molecular biology-related web sites.

To obtain further information on secondary structure prediction methods consult the molecular biology servers at the Swiss Institute of Bioinformatics (http://www.expasy.org) or at the EBML (http://www.embl-heidelberg.de).

4.3.7
Threading Methods

A most welcome situation in protein modeling occurs when, for a query protein, it is possible to find another protein that is highly homologous (30% or higher sequence similarity) and for which the structure has been already solved experimentally. In such cases, the above described classical comparative modeling approach allows for the construction of a protein model with sufficient accuracy. Another rather frequent situation arises when the sequence methods, or threading procedures can detect only weak similarity [72–76]. Consequently, the similarity of the unknown three-dimensional structure of the query sequence to the template structure cannot be quantified a priori. The two proteins may have identical topology; however, they may differ in their structurally not conserved regions. Also, particular secondary structure elements may be of different size, and there may be different packing between secondary structure elements. Frequently, the true structural similarity may be limited to only part of the structure having a common structural motif while the remainder of the protein has a completely different structure [76]. In this cases traditional comparative modeling methods fail and so-called fold recognition or threading methods have to be applied [72–78]. The earliest fold recognition methods were specifically designed to recognize folds in the absence of sequence similarity, and the sequence of the template protein was usually not taken into account at all. However, comparative modeling and threading approaches are nowadays often used simultaneously. Threading methods are closely related to the ab initio methods used for protein structure prediction [79], but whereas the ab initio methods have to explore all possible conformations, threading methods limit the search space to the conformations of known structures. Thus, threading methods fail for any protein which adopts a completely new fold.

The general threading approach involves taking a sequence and testing it on each member of a library of known protein structures. On each template, one must find the optimal sequence-to-structure alignment according to a certain score or force field. These alignments are then ranked on the calculated scores and the best ones are considered reliable candidates [72, 80]. A collection of often applied threading methods can be found in the literature [72–76].

A variety of different scoring functions have been developed for threading [72, 73, 80, 81] but most of them share some common properties. The functions applied have to bee simple since threading calculations often require a large number of possibilities to be considered. Many of the scoring functions used in threading programs are potentials of mean force which are also called knowledge-based potentials [80, 82].

They are quite different from the traditional force fields (which were described in general in section 2.2.1). The basic idea of knowledge-based force fields is that molecular structures observed from X-ray analysis or NMR contain a wealth of information on the stabilizing forces within macromolecules. Using statistical methods, the underlying rules governing the 3D structure of proteins have been revealed. It is the basic assumption of the Boltzmann principle that frequently observed states correspond to low-energy states of a system. Thus, the potentials of mean force are compiled by extracting relative frequencies of particular atom pair interactions from a database of protein structures [82]. The potentials of mean force consist usually of interactions among particular atom pairs and protein–solvent interactions. They incorporate all kinds of forces (electrostatic, dispersion, etc.) acting between particular protein atoms as well as the influence of the surrounding solvent on the interaction and can therefore be used to predict the structure of a macromolecule from its primary sequence. Potentials of mean force have been applied for the prediction of protein folds and even for the detection of errors in protein models and experimentally determined structures [80–82].

The utility of a protein model, either generated by comparative modeling or threading, depends upon the use to which the model is put. The accuracy of ab initio and threading models is too low for problems requiring high-resolution structural information, such as the traditional structure-based drug design. Instead, the low-resolution model produced by these methods can reveal structural and functional relationships between proteins not apparent from their amino acid sequence and provide a framework for analyzing spatial relationships between evolutionary conserved residues or between residues shown experimentally to be functionally important. To evaluate the different approaches for protein structure prediction, a competition called "Critical Assessment of techniques for protein Structure Prediction (CASP)" was organized in 1994/95 [83]. In these CASP experiments the scientific community was invited to predict the three-dimensional structures of novel proteins from their amino acid sequences. The three-dimensional structures of the proteins were solved by X-ray crystallography but were not published. Until now four CASP competitions have taken place. During CASP4 it was concluded that modeling of protein structures has now matured into a practical technology. It is possible, in principle, to produce useful models for more than half of the sequences entering the general sequence databases [79]. The CASP competitions provide a solid basis for assessing the reliability of protein models and their underlying modeling approaches.

References

[1] Blundell, T. L., and Johnson, L. N. *Protein Crystallography*, Academic Press: New York 1976.

[2] Bashford, D., Chothia, C., and Lesk, A. M. *J. Mol. Biol.* **196**, 199–216 (1987).

[3] Greer, J. *J. Mol. Biol.* **153**, 1027–1042 (1981).

[4] Chothia, C., and Lesk., A. M. *J. Mol. Biol.* **160**, 309–342 (1982).

[5] Johnson, M. S., Srinivasan, N., Sowdhamini, R., and Blundell, T. L. *Crit. Rev. Biochem. Mol. Biol.* **29**, 193–316 (1994).

[6] Sali, A., Overington, J. P., Johnson, M. S., and Blundell, T. L. *TIBS* **15**, 235–240 (1990).

[7] Jones, T. A., and Thirup, S. *EMBO J.* **5**, 819–822 (1986).

[8] Dudek, M. J., and Scheraga, H. A. *J. Comput. Chem.* **11**, 121–151 (1990).

[9] Levin, R. *Science* **237**, 1570 (1987).

[10] Thornton, J. M., and Gardner, S. P. *TIBS* **14**, 300–304 (1989).

[11] Orengo, C. A., Brown, N. P., and Taylor, W. R. *Proteins Struct. Func. Gen.* **14**, 139–146 (1992).

[12] Pearson, W. R. *Methods in Enzymology* **183**, 63–98 (1990).

[13] Altschul, S. F., Gish, W., Miller, W., Myers, E. W., and Lipman, D. J. *J. Mol. Biol.* **215**, 403–410 (1990).

[14] HOMOLOGY and MODELLER, Accelrys, San Diego, U.S.A.

[15] Sali, A., and Blundell, T. L. *J. Mol. Biol.* **234**, 779–815 (1993).

[16] SYBYL BIOPOLYMER, Tripos Associates, St. Louis, Missouri, U.S.A.

[17] WHAT IF, Vriend, G., European Molecular Biology Laboratory (EMBL), Heidelberg, Germany.

[18] Devereux, J., Haeberli, P., and Smithies, O. *Nucleic Acids Res.* **12**, 387–395 (1984).

[19] Barton, G. J. *Methods Enzymol.* **183**, 403–428 (1990).

[20] Needleman, S. B., and Wunsch, C. D. *J. Mol. Biol.* **48**, 443–453 (1970).

[21] Sander, C., Schneider, R. *Proteins Struct. Func. Gen.* **9**, 56–58 (1991).

[22] Schuler, G. D., Altschul, S. F., and Lipman, D. J. *Proteins Struct. Func Gen.* **9**, 180–190 (1991).

[23] Vingron, M., and Argos, P. *Comput. Appl. Biosci.* **5**, 115–121 (1989).

[24] Bowsell, D. R., and McLachlan, A. D. *Nucleic Acids Res.* **12**, 457–465 (1984).

[25] Dayhoff, M. O., Schwartz, R. M., and Orcutt, B. C. A Model of Evolutionary Change in Proteins. In: *Atlas of Protein Sequence and Structure*, Vol. 5, Suppl. 3. Dayhoff, M. O. (Ed.). Natl. Biomed. Res. Found.: Washington; 345–352 (1978).

[26] Gribskov, M., McLachlan, A. D., and Eisenberg, D. *Proc. Natl. Acad. Sci. U.S.A.* **84**, 4355–4358 (1987).

[27] Risler, J. L., Delorme, M. O., Delacroix, H., and Henaut, A. *J. Mol. Biol.* **204**, 1019–1029 (1988).

[28] Engelman, D. M., Steitz, T. A., and Goldman, A. *Anu. Rev. Biophys. Chem.* **15**, 321 (1986).

[29] Gonnet, G. H., Cohen, M. A., and Benner, S. A. *Science* **256**, 1443–1445 (1992).

[30] Doolittle, R. *Methods Enzymol.* **183**, 736–772 (1990).

[31] Barton G. J., and Sternberg, M. J. E. *J. Mol. Biol.* **198**, 327–337 (1987).

[32] Chothia, C., and Lesk, A. M. *EMBO J.* **5**, 823–826 (1986).

[33] Perutz, M. F., Kendrew, J. C., and Watson, H. C. *J. Mol. Biol.* **13**, 669–678 (1965).

[34] Kabsch, W. *Acta. Cryst.* **A32**, 922–923 (1976).

[35] Kabsch, W. *Acta. Cryst.* **A34**, 827–838 (1978).

[36] McLachlan, A. D. *Acta. Cryst.* **A38**, 871–873 (1982).

[37] Taylor, W. R. *J. Mol. Biol.* **188**, 233–258 (1986).

[38] Crippen, G. M., and Havel, T. F. *Acta. Cryst.* **A34**, 282–284 (1978).

[39] Vriend, G., and Sander, C. *Proteins Struct. Func. Gen.* **11**, 52–58 (1991).

[40] Matthews, B. W., and Rossmann, M. G. *Methods Enzymol.* **115**, 397–420 (1985).

[41] Kabsch, W., and Sander, C. *Biopolymers* **22**, 2577–2637 (1983).

[42] Frishman, D., and Argos, P. *Proteins Struct. Func Gen.* **23**, 566–579 (1995).

[43] Sali , A., *Curr. Opin. Biotechnology* **6**, 437–451 (1995).

[44] Bruccoleri, R. E., and Karplus, M. *Biopolymers* **26**, 137–168 (1987).

[45] Bruccoleri, R. E., Haber, E., and Novotny, J. *Nature* **335**, 564–568 (1988).

[46] Shenkin, P. S., Yarmush, D. L., Fine, R. M., Wang, H., and Levinthal, C. *Biopolymers* **26**, 2053–2085 (1988).

[47] Feldmann, R. J., Bing, D. H., Potter, M., Mainhart, C., Furie, B., Furie, B. C., and

Caporale, L. H. *Ann N. Y. Acad. Sci.* **439**, 12–43 (1985).

[48] Blundell, T. L., Sibanda, B. L., and Pearl, L. *Nature* **304**, 273–275 (1983).

[49] Palmer, K. A., Scheraga, H. A., Riordan, J. F., and Vallee, B. L. *Proc. Natl. Acad. Sci. U.S.A.* **83**, 1965–1969 (1986).

[50] Ponder, J., and Richards, F. M. *J. Mol. Biol.* **193**, 775–791 (1987).

[51] Summers, N. L., Carlson, W. D., and Karplus, M. *J. Mol. Biol.* **196**, 175–198 (1987).

[52] McGregor, M. J., Islam, S. A., and Sternberg, M. J. *J. Mol. Biol.* **198**, 195–210 (1987).

[53] Benedetti, E., Morelli, G., Nemethy, G., and Scheraga, H. A. *Int. J. Peptide Protein Res.* **22**, 1–15 (1983).

[54] Sutcliffe, M. J., Hayes, F. R. F., and Blundell, T. L. *Protein Eng.* **1**, 385–392 (1987).

[55] Schrauber, H., Eisenhaber, F., and Argos, P. *J. Mol. Biol.* **230**, 592–612 (1993).

[56] Dunbrack, R. L. Jr., and Karplus, M. *J. Mol. Biol.* **230**, 543–574 (1993).

[57] Havel, T. F., and Snow, M. *J. Mol. Biol.* **217**, 1–7 (1991).

[58] Srinivasan, S., March., C. J., and Sudarsanam, S. *Protein Science* **2**, 277–289 (1993).

[59] Havel, T. F. *Molecular Simulations* **10**, 175–210 (1993).

[60] Fasman, G. D. *Trends Biochem. Sci.* **14**, 295–299 (1989).

[61] Chou, P. Y., and Fasman, G. D. *Biochemistry* **13**, 211–245 (1974).

[62] Garnier, J., Osguthorpe, D. J., and Robson, B. *J. Mol. Biol.* **120**, 97–120 (1978).

[63] Biou, V., Gibrat, J. F., Levin, J. M., Robson, B., and Garnier, J. *Protein Eng.* **2**, 185–191 (1988).

[64] Lim, V. I. *J. Mol. Biol.* **88**, 873–894 (1974).

[65] Eisenberg, D., Weiss, R. M., and Terwilliger, T. C. *Nature* **299**, 371–374 (1982).

[66] Kyte, J., and Doolittle, R. F. *J. Mol. Biol.* **157**, 105–132 (1982).

[67] Rost, B., and Sander, C. *J. Mol. Biol.* **232**, 584–599 (1993).

[68] Rost. B., and Sander, C. *Proteins Struct. Func. Gen* **19**, 55–72 (1994).

[69] Rost, B., and Eyrich, V. A. *Proteins Struct. Func. Gen. Suppl.* **5**, 192–199 (2001).

[70] McGuffin, L. J., Bryson, K., and Jones, D. T. *Bioinformatics* **16**, 404–405 (2000).

[71] Cuff, J. A., Clamp, M. E., Siddiqui, A. S., Finlay, M., and Barton, G. J. *Bioinformatics* **14**, 892–893 (1998).

[72] Bryant, S. H. *Proteins Struct. Func. Gen.* **26**, 172–185 (1996).

[73] Jones, D. T. *J. Mol. Biol.* **287**, 797–815 (1999).

[74] Wilmanns, M., and Eisenberg, D. *Protein Eng.* **8**, 626–635 (1995).

[75] Skolnick, J., and Kihara, D. *Proteins Struct. Func. Gen.* **42**, 319–331 (2001).

[76] Panchenko, A. R., Marchler-Bauer, A., and Bryant, S. H. *J. Mol. Biol.* **296**, 1319–1331 (2000).

[77] Kolinski, A., Betancourt, M. R., Kihara, D., Rotkiewicz, P., and Skolnick, J. *Proteins Struct. Func. Gen.* **44**, 133–149 (2001).

[78] Xu, Y., and Xu, D. *Proteins Struct. Func. Gen.* **40**, 343–354 (2000).

[79] Moult, J. *Curr. Opin. Biotechnology* **10**, 583–588 (1999).

[80] Sippl, M. J. *J. Mol. Biol.* **213**, 859–883 (1990).

[81] Jones, D. T., and Thornton, J. M. *Curr. Opin. Struct. Biology* **6**, 210–216 (1996).

[82] Sippl, M. J. *Proteins Struct. Func. Gen.* **17**, 355–362 (1993).

[83] Mosimann, S., Meleshko, S., and Jones, M. N. G. *Proteins Struct. Func. Gen.* **23**, 301–317 (1995).

4.4

Optimization Procedures—Model Refinement—Molecular Dynamics

4.4.1

Force Fields for Protein Modeling

Protein models derived from either comparative modeling or crystal structures need further refinement. In the course of generating protein models the loop and side chain conformations in general are chosen arbitrarily; therefore the conformations do not correspond to energetically reasonable structures. Also crystal structures must be relaxed in order to remove the internal strain resulting from the crystal packing forces or to remove close contacts between hydrogen atoms or amino acid residues which may have been added to the crystal coordinates after structure determination.

As protein models consist of hundreds or thousands of atoms the only feasible methods of computing systems of such size are molecular mechanics calculations. The common force fields used in molecular mechanics calculations are based in principle on the equations for the potential energy function as described in section 2.2.1. However, force fields for protein modeling differ in some respect from small molecule force fields. Besides the specific parametrization for proteins and DNA, certain simplifications are frequently introduced. In some force fields non-polar hydrogens are not represented explicitly, but are included into the description of the heavy atoms to which they are bonded. In contrast, polar hydrogens which may act as potential partners in hydrogen bonding are treated explicitly. This procedure is denoted as the *united atom model*. In the AMBER [1, 2] force field both the united-atom model or an all-atom representation can be applied, while the GROMOS force field [3] offers only the united atom model. Other simplifications can be made by introducing cut-off radii [4] to reduce the time-consuming part of calculating non-bonded interactions between atoms separated by distances larger than a defined cut-off value.

An additional variation is made in respect of the treatment of the electrostatic interactions. As the explicit inclusion of solvent is still a problem, some force fields try to simulate the solvent effect by introducing a distance-dependent dielectric constant [1, 2]. Especially in the case of macromolecules the electrostatic field in the environment of the system can not be considered to be continous. Thus, a differentiating procedure in calculating the particular properties is necessary in order to reflect the electrostatic effects which depend on the local situation, e.g. in the binding pocket or on the surface of the protein. A detailed discussion of this subject and a description of methods handling the complex situation adequately is given in section 4.6.1.

The modifications established in protein force fields are various and can not be discussed here in detail. A comprehensive description of potential simplifications is given in [5]. It should be borne in mind that each simplification applied can result in a loss of accuracy. The decision on the force field to be chosen strongly depends on the problem to be investigated; hence the most accurate force field which is applicable for the whole study must always be selected. The use of different force fields within a molecular modeling study should generally be avoided.

There are several common force fields for protein modeling implemented in software programs. The following list is not complete but comprises some of the most frequently employed methods: AMBER [1, 2], CVFF [6], CHARMM [7] and GROMOS [3].

4.4.2
Geometry Optimization

The algorithms used in the minimization procedures for proteins are the same as for small molecules and have been discussed in detail in section 2.2.3. The minimization algorithms applied to optimize the geometry usually find only the local minimum on the potential energy surface closest to the initial coordinates. In case of a well-resolved crystal structure the minimization will directly yield one energetically favorable conformation. The relaxation of a crystal structure usually is a straightforward procedure. However, crystal coordinates—even if highly resolved—sometimes have several unfavorable atomic interactions. These disordered atomic positions cause large initial forces that result in artificial movements away from the original structure when starting the minimization process. A general approach to avoid these large deviations is to relax the protein model gradually.

A more profound solution would be to assign tethering forces to all heavy atoms of the crystal structure in the first stage of minimization. The *tethering constant* is a force applied to fix atomic coordinates on predefined positions. The strength of the tethering force can be selected by the user and affects the extent of movement of the atoms measured by the rms deviation from the initial coordinates. When tethering the heavy atoms the hydrogen atoms, and perhaps solvent molecules, are allowed to adjust their positions in order to minimize the total potential energy. A suitable minimization method for this purpose is the steepest descent algorithm. For this initial relaxation step a crude convergence criterion can be applied or the process can be finished by defining a maximum number of allowed minimization steps.

Subsequently it is recommended to tether only the well-defined main chain atoms. Now the side chains are allowed to move and to adjust their orientations. The steepest descent method is suitable also in this case. Ultimately the restraints are removed in the last step so that the final minimum represents a totally relaxed conformation. The minimization algorithm should be changed to conjugate gradient to reach convergence in an effective way.

The application of tethering forces can also be useful and necessary in the modeling of incomplete systems. These may result in an X-ray study if certain parts of the crystals or included solvent molecules cannot be resolved adequately. Also active site models of enzymes or binding pockets of proteins used for the investigation of potential ligand–protein interactions are examples of typical incomplete systems.

Due to the absence of neighboring amino acids or solvent molecules the atom positions at the surface of a protein are mobile. As a consequence, large deviations from the initial positions will result after minimization and the final geometry must be regarded as an artefact. Therefore atoms or the ends of side chains are tethered at their original positions to avoid unrealistic atom movements at the surface of the protein.

With the objective to confirm the accuracy of the relaxed protein model the deviations from the experimental structure should be examined. For this purpose the initial structure and the final geometry are superimposed using least-squares fit methods. Normally either all backbone atoms or only backbone atoms of the well-refined secondary structural elements are used as fitting points. The quality of the fit can be judged by the rms deviation of the optimized form from the initial geometry. The value of the rms deviation is strongly dependent on the number and localization of atoms which are considered for the fit. Naturally, a fit of all heavy atoms would result in a much higher rms value than a fit which is confined to backbone atoms only, mainly due to the greater mobility of side chains.

If the generated model is based merely on comparative modeling techniques the loop and side chain conformations need further refinement. It is necessary to investigate carefully their conformational behavior and to analyze the potential energy surface for other possible low-energy conformations. A valuable tool for this purpose are *molecular dynamics simulations*. The relaxed geometry obtained as result of a minimization procedure can be used as starting point for molecular dynamics simulations.

4.4.3
The Use of Molecular Dynamics Simulations in Model Refinement

As mentioned above the refinement of models derived from comparative modeling is a must. Loop and side chain conformations of the derived protein model represent only one possible conformation and the minimum structure found by the minimization algorithms represents only one local minimum. In order to detect the energetically most favored 3D structure of a system a modified strategy is needed for searching the conformational space more thoroughly.

Molecular dynamics simulations offer an effective means to solve this problem, especially for molecules containing hundreds of rotatable bonds. A molecular dynamics simulation is performed by integrating the classical equations of motion over a period of time for the molecular system. The resulting trajectory for the molecule can be used to compute the average and time-dependent properties of the system. The theory of the molecular dynamics method and its application in conformational searching of small molecules have been discussed and illustrated on some impressive examples in section 2.3.3. Here, we will focus on the utilization of this technique in the refinement of 3D macromolecular structures.

The use of molecular dynamics has made an essential contribution to the understanding of dynamic processes in proteins at the atomic level. However, there are some basic limitations and problems arising with increasing size and associated with the immense number of degrees of freedom of large molecular systems.

Although computer resources have become sufficiently powerful to enable handling of quite large systems it is still necessary to introduce some modifications in order to reduce the demanded computation time [5]. A very useful side effect of the simplifications employed is the fact that they open the possibility of longer time periods to be chosen for the sampling of the dynamic simulation. This offers a way of observing the dynamic behavior of large molecular systems more completely.

Before discussing the various possibilities in detail it must be mentioned again that each modification and reduction of the number of degrees of freedom can cause a lack of accuracy and it has to be checked carefully whether or not a respective simplification can be tolerated.

One basic and very common simplified procedure is the use of *united atom potential energy functions*. The underlying theory of this methodology has been described earlier. Most of the force fields for protein modeling, such as AMBER [1, 2] and GROMOS [3] are based on these algorithms. Omission of the non-polar hydrogens in a united-atom force field does significantly reduce the number of particles in a large biomolecule. A further possibility to reduce the demand for computer time is provided by application of the SHAKE [8] algorithm. In the SHAKE procedure additional forces are assigned to the atoms, aiming to keep bond lengths fixed at equilibrium values. This is very useful for several reasons. Above all, bond stretching energy terms must not be calculated for the frozen bonds. The magnitude of the integration step depends on the fastest occurring vibrations in a molecule. This is usually the high frequency vibration of the C–H bond stretching. This period is of the order of 10^{-14} seconds; therefore the integration step should be chosen to be 10^{-15} seconds (1femtosecond). Applying the SHAKE algorithm to this type of C–H bond allows a larger integration step with the effect of reducing the necessary computational expense, and thereby offering the chance of simulating the system over a longer time period. The definition of cut-off radii, leading to a neglect of non-bonded interactions beyond the defined distance, also yields the same effect.

In addition, the application of a well-balanced computational protocol may save computer time. In this respect several parts of a protein can be kept rigid and the molecular dynamics simulations then carried out only for flexible parts such as loops or side chains, while well-defined secondary structures like α-helices or β-strands in the core of the protein are not taken into account. The availability of NMR data can also be a reason to fix atoms, side chains or parts of the protein at their initial coordinates in order to impede their movement away from the experimentally derived positions. Again, a warning must be given; restraining parts of flexible molecules leads to a reduction in the number of degrees of freedom. Without any doubt a more comprehensive exploration of the conformational space, and hence better results, are achieved when no positional restraints are applied on parts of the protein structure.

All mentioned methods enhance the efficiency of the molecular dynamics simulations; nevertheless, for some problems the feasible time scale is still too short. If for example the binding of a ligand to an enzyme or receptor protein—as well as the thereby triggered conformational change—is to be studied, the time required for this process can be in the order of picoseconds or even nanoseconds [9]. The same time-scale would be indispensable for a simulation of protein folding. Both types of problems are still out of reach.

References

[1] Weiner, S. J., Kollman, P. A., Case, D. A., Singh, U. C., Ghio,C., Alagona, G., Profeta, S. Jr., and Weiner, P. *J. Am. Chem. Soc.* **106**, 765–784 (1984).

[2] Weiner, S. J., Kollman, P. A., Nguyen, D. T., and Case, D. A. *J. Comput. Chem.* **7**, 230–252 (1986).

[3] van Gunsteren, W. F., and Berendsen, H. J. C. Molecular dynamics simulations: techniques and applications to proteins. In: *Molecular Dynamics and Protein Structure*. Hermans, J. (Ed.). Polycrystal Books Service: Western Springs; 5–14 (1985).

[4] Brooks, C. L., III, Montgomery, Pettitt, B., and Karplus, M. *J. Chem. Phys.* **83**, 5897–5908 (1985).

[5] van Gunsteren, W. F. *Adv. Biomol. Simul.* **239**, 131–146 (1992).

[6] Dauber-Osguthorpe, P., Roberts, V. A., Osguthorpe, D. J., Wolff, J., Genest, M., and Hagler, A.T. *Proteins Struct. Func. Gen.* **4**, 31–47 (1988).

[7] Brooks, B. R., Bruccoleri, R. E., Olafson, B. D., States, D. J., Swaminathan, S., and Karplus, M. *J. Comput. Chem.* **4**, 187 (1983).

[8] Ryckaert, J. P., Ciccotti, G., and Berendsen, H. J. C. *J. Comput. Phys.* **23**, 327 (1977).

[9] Lybrand, T. P. Computer Simulation of Biomolecular Systems Using Molecular Dynamics and Free Energy Perturbation Methods. In: *Reviews in Computational Chemistry*, Vol. 1. Lipkowitz, K. B., and Boyd, D. B. (Eds.). VCH: New York; 295–320 (1990).

[10] Auffinger, P., and Wipff, G. *J. Comput. Chem.* **11**, 19–31 (1990).

[11] Kirkpatrick, S., Gelatt, C. D., and Vecchi, M. P. *Science* **220**, 671–680 (1983).

[12] Salvino, J. M., Seoane, P. R., and Dolle, R. E. *J. Comput. Chem.* **14**, 438–444 (1993).

[13] Wilson, S. R., and Cui, W. *Biopolymers* **29**, 225–235 (1990).

[14] Mackey, D. H. J., Cross, A. J., and Hagler, A. T. The Role of Energy Minimization in Simulation Strategies of Biomolecular Systems. In: *Prediction of Protein Structure and the Principles of Protein Conformation*. Fasman, G. (Ed.). Plenum Press: New York; 317–358 (1989).

[15] Kerr, I. D., Sankararamakrishnan, R., Smart, O. S., and Sansom, M. S. P. *Biophys. J.* **67**, 1501–1515 (1994).

[16] Bruccoleri, R. E., and Karplus, M. *Biopolymers* **29**, 1847–1862 (1990).

[17] Vijayakumar, S., Ravishanker, G., Pratt, R. F., and Beveridge, D. L. *J. Am. Chem. Soc.* **117**, 1722–1730 (1995).

[18] van Gunsteren, W. F., and Karplus, M. *Biochemistry* **21**, 2259–2274 (1982).

[19] Karplus, M., and Mc Cammon, J. A. *Nat. Struct. Biol.* **9**, 646–652 (2002).

4.5
Validation of Protein Models

Once a protein model has been built using comparative modeling methods and subsequently optimized by molecular mechanics or molecular dynamics, it is important to assess its quality and reliability. The question arises how protein models can be tested for correctness and accuracy. This is a very difficult business, because the quality of a homology-based protein model depends on a huge number of properties on different levels of structural organization. This is summarized in Fig. 4.10.

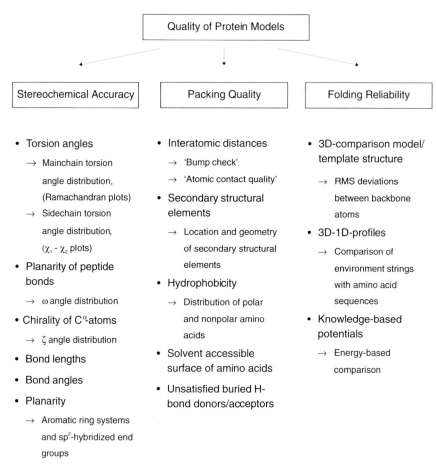

Figure 4.10 Quality questionnaire for protein models.

4.5.1
Stereochemical Accuracy

The quality of the 3D structure of a protein model depends strongly on the accuracy of the used template structure, i.e. the quality of the crystal structure [1]. Of course, the modeled protein cannot show higher accuracy than the crystal structure which has been used as a template. Protein structures derived from X-ray diffraction can contain errors, both experimental and in the interpretation of the results [1–3]. The general measures for the quality of crystal structures are the resolution and the R-factor. The better the resolution of the protein crystal the greater the number of independent experimental observations derived from the diffraction data and hence the greater the accuracy of the protein structure [4]. The resolution of protein structures contained in the Protein Data Bank is usually found to be in the range of 1–4 Å. The R-factor is a measure for the agreement between the derived 3D structure of a protein crystal (the 3D structure which fits the electron density map best) and the "real" crystal structure. The R-factor can be determined by comparing the experimentally obtained amplitudes of the X-ray reflections and the amplitudes calculated from the protein structure which shows the best fit to the electron density map (for a detailed discussion about the accuracy of protein X-ray crystallography the reader is referrred to the literature [5]). The better the agreement between observed and calculated amplitudes (resulting in a low R-factor), the better the agreement between the derived and the real crystal structure. The R-factor can be artificially reduced in a number of ways and therefore sometimes might be misleading [2]. It is commonly accepted to consider structures with a resolution of 2.0 Å or better to be reliable. If in addition the R-factor is below 20% it can be safely assumed that the protein structure is essentially correct.

To verify the stereochemical quality of a model-built structure, the accuracy of parameters such as bond lengths, bond angles, torsion angles and correctness of the amino acid chirality, must be proved. It has been observed in 3D structures of proteins that mainly the bond lengths and angles cluster around the "ideal values". Thus, the mean values detected in crystal structures can be regarded as good indicators of the stereochemical quality and must be compared with the actual values in the generated protein model (see Table 4.1) [6] in order to discover stereochemical irregularities which would disclose a bad structure.

Since a manual inspection of all stereochemical parameters of a protein will be tedious and time-consuming, programs have been developed which automatically check all stereochemical properties. Examples are PROCHECK [7], WHAT CHECK [8] and PROVE [9] which are accessible through the EMBL homepage (http://www.biotech.embl-heidelberg.de).

One important indicator of stereochemical quality is the distribution of the main chain torsion angles ϕ and ψ. The distribution of all ϕ and ψ torsion angles in a protein can be examined in a Ramachandran plot. As described in section 4.2.1, the favored and unfavored regions of the classical Ramachandran plot have been determined by studying the conformational behavior of isolated dipeptides. Very conveniently the ϕ–ψ torsion angles observed in hundreds of well-refined protein structures generally lie within the same regions as determined for the isolated dipeptides. It is one of the remarkable properties of repetitive secondary structures in proteins that the observed ϕ,ψ-values are very close to the optimal dipeptide conformations, as calculated by Ramachandran. Also the ϕ and ψ torsion angles of non-repetitive structures, like loops or turns, are found within the favored regions of the Ramachandran plot, but are more widely distributed over these areas.

Table 4.1 Stereochemical parameters derived from high-resolution protein structures after Morris et al. [6]

Stereochemical parameters	Mean Value	Standard deviation
$\phi-\psi$ in most favored regions of Ramachandran plots	> 90%	–
χ_1 torsion angle gauche minus	64.1°	15.7°
trans	183.6°	16.8°
gauche plus	−66.7°	15.0°
χ_2 torsion angle	177.4°	18.5°
Proline ϕ torsion angle	−65.4°	11.2°
α-Helix ϕ torsion angle	−65.3°	11.9°
α-Helix ψ torsion angle	−39.4°	11.3°
Disulfide bond separation	2.0 Å	0.1 Å
ω Torsion angle	180.0°	5.8°
C^α tetrahedral distortion: z torsion angle (virtual torsion angle C^α–N–C–C^β)	33.9°	3.5°

As an example, the Ramachandran plot of a protein crystal structure (cephalosporinase of *enterobacter cloacae*) is shown in Fig. 4.11. The torsion angles of all residues, except those for proline residues and those at the chain termini, are presented. Glycine residues are separately identified by triangles (as those are not restricted to any particular region of the plot). The shading represents the different major regions of the plot: the darker the region the more favored is the corresponding ϕ,ψ combination. The white region is the disallowed region for normal amino acids and any residue found in this region must be carefully inspected. Usually amino acids lying in less-favored regions are especially labelled (in Fig. 4.11 shown in red) with residue name and residue number for easy indentification and inspection.

Unfavorable stereochemistry, becoming visible by disallowed ϕ,ψ torsion angles, seems to occur in natural proteins exclusively if the special geometry is required for function or stability, for example when residues in the core of the protein are involved in hydrogen bonds or salt-bridges. Residues which are allowed to lie outside the major regions of the Ramachandran plot are proline and glycine. Because glycine and proline have—due to their different stereochemistry—other favored and unfavored regions, it is more convenient to mark these amino acid types particularly or to exclude them from the normal Ramachandran plot. Therefore, separate Ramachandran plots for all glycines, all prolines, and all other amino acids are very often

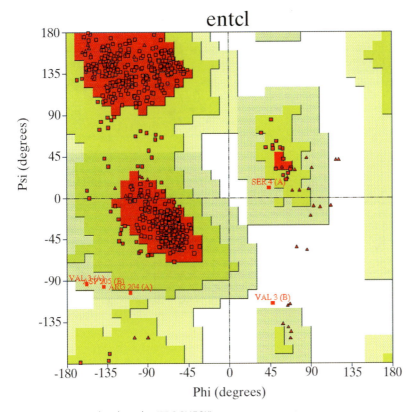

Figure 4.11 Ramachandran plot (PROCHECK).

created. The percentage of residues lying in the favored regions of a Ramachandran plot is one of the best guides to check stereochemical quality of a protein model. Ideally, one would hope to have more than 90% of the residues in the allowed regions [7].

The same check as described above for main chain torsion angles can be applied in case of the side chain torsion angles χ_i. The χ_1 torsion angles, observed in well-refined structures of proteins [6], are generally close to one of the three possible staggered conformations, the most favored conformation being the one where the bulkiest groups are most remote (see Table 4.1: gauche plus, trans, and gauche minus torsion angles for χ_1). For the χ_2 torsion angles a preference for the trans conformation has been found. A similar distribution for the side chain torsion angles in protein crystal structures has been detected by Ponder and Richards [10]. The distribution of the side chain torsion angles of all amino acid types in protein models can be inspected in more detail in graphs, where usually side chain torsion angles χ_1 are plotted versus χ_2. Examples for this kind of graph are shown in Fig. 4.12 for a cephalosporinase of *Enterobacter cloacae*. Every single plot shows the χ_1–χ_2 angle distribution for a particular amino acid type. The green shading on each plot indicates the favorable regions which have been determined from a data set of

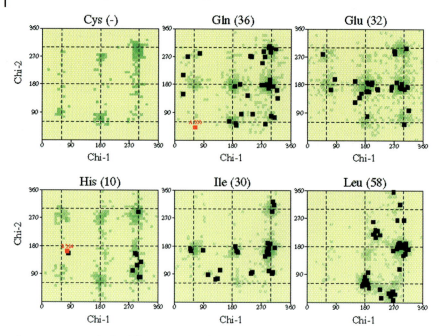

Figure 4.12 χ_1–χ_2 Plot for different amino acids (PROCHECK).

well-resolved protein crystal structures [7]. Black marks indicate the corresponding values found in the cephalosporinase; the red marks denote outliers.

Some of the stereochemical parameters of protein structure have been found to be constant in all known proteins. Of course, these properties are a very sensitive measure for the quality of protein models and must be carefully checked for consistency.

This list contains:

1. *The peptide bond planarity:* this property is usually measured by calculating the mean value and the standard deviation of all ω angles in the investigated protein. The smaller the standard deviation, the tighter the clustering around the normal value of 180°, which represents the planar trans configuration (see also Table 4.1 for the distribution of ω angles in crystal structures). All *cis* peptide bonds are also separately listed and must be inspected. *Cis* peptide bonds occur in proteins at about 5% of the bonds that precede proline residues. Regarding all peptide bonds, which do not involve proline residues, the *cis* configuration is observed less than 0.05% [11, 12].

2. *The chirality of the C^{α}-atoms:* One of the general principles of protein structure is the preference for one handedness over the other (e.g. the preference for the right-handed conformation of an α-helix). The basis for this is the presence of an asymmetric center at the C^{α}-atom which in all naturally occurring amino acids is L-configurated. A protein model must therefore be examined for correct chiralities. A parameter which provides a measure for the correct-

ness of chirality is the α torsion angle. This is a virtual torsion angle which is not defined by any actual bond in a protein. Rather this torsion angle is determined by the C^{α}–N–C'–C' atoms of each amino acid residue. The numerical values of the α torsion angle should reside between 23° and 45°. If the value is negative, this fact signifies the appearance of an incorrect D-amino acid [7].

3. *Main chain bond lengths and angles:* the distribution of each of the different main chain bond lengths and angles in a protein is compared with the distribution observed in well-resolved crystal structures. Usually deviations more than 0.05 Å for bond lengths and 10° for bond angles are regarded as distorted geometries which have to be inspected in detail [3].

Aromatic ring systems (Phe, Tyr, Trp, His) and sp^2-hybridized end groups (Arg, Asn, Asp, Glu, Gln) must be checked for planarity. The deviation of these parameters, i.e. the distorted geometry, is often the result of bad interatomic contacts. Removing the steric constraints and subsequently optimizing the model in most cases yields a relaxed structure with ideal geometrical parameters.

4.5.2
Packing Quality

Specific packing interactions within the interior are assumed to play an important role for the structural specifity of proteins [13–15]. It has been observed that globular proteins are tightly packed with packing densities comparable with those found in crystals of small organic molecules [13]. The interior of globular proteins contains side chains that fit together with striking complementarity, like pieces of a 3D jigsaw puzzle. The high packing densities observed in proteins are the consequence of the fact, that segments of secondary structure are packed together closely; helix against helix, helix against strands of a β-sheet, and strands against strands of different β-sheets [15–18]. The interior packing of globular proteins is a major contribution to the stability of the overall conformation. Therefore, the packing quality of a protein model can be used to estimate its reliability. It can be judged using a variety of methods, which will be described in detail in this section.

In a first step, it must be verified that the generated and refined protein model includes no bad van der Waals contacts. Therefore, all interatomic distances must be examined for residing in ranges which have been observed in well-refined crystal structures. Several procedures exist for this distance check. In the simplest, all interatomic distances are measured and those with distances below a determined threshold are defined as bad contacts which have to be inspected in detail (for example, 2.6 Å is used as threshold in the PROCHECK program [7]). A more accurate judgement of interatomic distances is performed by programs like WHATCHECK [8]. For all well-refined protein crystal structures stored in the Protein Data Bank all interatomic distances shorter than the sum of their van der Waals radii +1.0 Å are determined and stored. The distance that subdivides the collected values such that 5% of all observed distances are shorter and 95% are longer than this measure, is defined as "short normal distance". As there are 163 different atom types in the naturally occuring amino acids 163 × 163 "short normal distances" are defined. All distances occuring in the protein model which are more than 0.25 Å shorter than the short normal distances are reported by the program.

The next step involves the examination of the secondary structural elements of the protein model. As we have already mentioned in section 4.3.2, the secondary structural elements are the most conserved regions in highly homologous proteins. Thus, it must be proven whether the secondary structural elements observed in the template protein also can be detected in the protein model, i.e. whether the secondary structure has been maintained during the building and optimization process. Programs which can be applied for this purpose are the DSSP [19] or the STRIDE program [20] (see section 4.3.2). These programs allow a more sophisticated assignment of secondary structure than the manual inspection of α-helices and β-sheets.

A variety of methods exist, which use the huge amount of information derived from protein crystal structures to estimate the packing quality of model built structures [21–24]. From the assumption that atom–atom interactions are the primary determinant of protein conformation, Vriend and Sander have developed a program that checks the packing quality of a protein model by calculating a so-called "contact quality index" [21]. This index is a measure of the agreement between the distributions of atoms around an amino acid side chain in the protein model and equivalent distributions observed in well-resolved protein structures. For that reason a database has been generated which contains a contact probability distribution for all amino acid side chains. This magnitude describes the probability for a certain atom type to occur in a particular region around the side chain. These probability values are used to check the contact quality in the protein model. The better the agreement between the distributions in the model and in the crystal structures the higher the contact quality index, and the more favorable the residue packing.

The distribution of polar and non-polar residues between the interior and the surface of proteins has been found to be a general principle of the architecture of globular proteins. At a simple level, a globular protein can be considered to consist of a hydrophobic interior surrounded by a hydrophilic external surface which interacts with the solvent molecules. These building principles have been identified in most 3D structures of globular proteins and can be summarized as follows:

1. The interior of globular proteins is densely packed without large empty space and is generally hydrophobic. Non-polar side chains predominate in the protein interior; Val, Leu, Ile, Phe, Ala, and Gly residues comprise 63% of the interior amino acids [11]. Ionized pairs of acidic and basic groups hardly occur in the interior, even though such pairs might be expected to have no net charge due to the formation of salt-bridges.

2. Charged and polar groups are located on the surface of globular proteins accessible to the solvent. On average, Asp, Glu, Lys, and Arg residues comprise 27% of the protein surface and only 4% of the interior residues [11]. (Integral membrane proteins differ from globular proteins primarily in having extremely non-polar surfaces which are in contact with the hydrophobic membrane core.)

These features make a major contribution to the stability of folded proteins [15, 25, 26]. The underlying principle for this distribution is the hydrophobic effect, i.e. the removal of hydrophobic residues from contact with water. It has been observed that the free energies, associated with the transfer from water to organic solvent, of polar, neutral and non-polar residues are correlated with the extent to which they occur in the interior and exterior of proteins [27]. Therefore, the distribution of hydrophobic and hydrophilic residues in proteins, can be used to estimate the reliability of protein models [27–30]. Several programs have been developed which use this feature as a measure of the packing quality [8, 29, 30] of a protein model.

It has been also observed that the hydrophobicity of an amino acid (defined as free energy of transfer from water to organic solvent) is related linearly to its surface area, i.e. the more hydrophobic the residue, the more completely buried it will be [31]. The buried surface area of a particular amino acid is herein defined as the difference between the solvent-accessible surface of the residue in an extented polypeptide chain (usually defined as the "standard state" in the tri-peptide Gly-XXX-Gly) and the solvent-accessible surface of the residue in the folded protein. It has been demonstrated that the buried surface area, i.e. the area which is lost when a residue is transferred from the defined "standard state" to a folded protein, is proportional to its hydrophobicity.

Additionally, the total surface buried within globular proteins has been found to correlate with their molecular weights, i.e. upon folding, globular proteins bury a constant fraction of their available surface [27]. Several programs have been developed which use the general properties of amino acid surfaces in order to provide an estimation of the packing quality of globular proteins [8, 29, 31]. For a detailed review of the topic of molecular surfaces and their contributions to protein stability the reader is referred to the literature [15, 32].

Although the residues that form the protein interior are usually non-polar or neutral, there are rare cases of buried polar residues. It has been observed in many investigations of protein crystal structures that virtually all polar groups in the protein interior are paired in hydrogen bonds. Many of these polar groups form hydrogen bonds within their own secondary structure (i.e. α-helices and β-sheets). Others are involved in binding co-factors, metal ions or are located in the active site of proteins. Buried ionizable groups, which rarely occur inside globular proteins, are usually always involved in salt-bridges. Sometimes the positive and negative charges are bridged by water molecules. Due to this observation, it is necessary to check the protein model whether all polar buried residues are paired in hydrogen bonds and whether all charged residues are involved in salt-bridges. Salt-bridges and hydrogen bonds are usually identified on the basis of their interatomic distances [33].

4.5.3
Folding Reliability

Proteins with homologous amino acid sequences generally have similar folds. Therefore the overall 3D structure of the protein model and its template should be similar. Especially in the structurally conserved regions, the homologous proteins must possess the same conformation. In cases where the originally constructed protein model contains large regions of steric strain (due to the incorrect architecture), the protein may undergo correspondingly large movements in its 3D structure during the refinement process. The resulting protein conformation is not reliable, because it shows only little agreement with the 3D structure of the template protein. When checking protein conformations, one normally measures the similarity in 3D structure by the rms deviations of the C^α-atomic or the backbone coordinates after optimal rigid body superposition of the two structures (for details, see [34]). A very large rms deviation means the two structures are dissimilar; a value of zero means that they are identical in conformation. Homologous proteins generally show low rms deviations for their C^α-atoms, but no general value exists which can be used as an indicator whether two protein structures are similar or dissimilar. Chothia and Lesk [35] have performed an investigation on structural similarity of homologous proteins. The overall extent of the structural divergence of two homologous proteins was measured by optimally superposing the common conserved regions (the so-called common core) and calculating the rms difference in the positions of their backbone atoms. For a test set of 32 homologous pairs of proteins they have found rms differences for the common cores which vary between 0.62 and 2.31 Å (see Fig. 4.13).

When the overall structural similarity of the protein model and the template protein have been evaluated, the question arises whether the generated conformation for an unknown protein is the correct native fold. How can one prove whether the constructed model is correct in its overall conformation? In the search for criteria that discriminate between the correct conformation and incorrectly folded models Novotny et al. have performed an interesting investigation [36]. They have studied

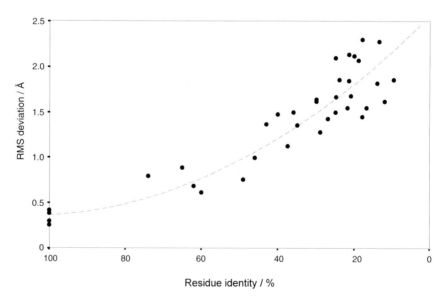

Figure 4.13   rms/sequence identity plot.

two structurally dissimilar but identically large proteins, hemerythrin (1HMQ) and the variable domain of mouse immunoglobulin κ-chain (1MCP-L). The two proteins have been modified by placing the amino acid sequence of one protein on to the backbone structure of the other and vice versa, in order to obtain incorrect models. The model structures were optimized to remove steric overlaps of side chains. After minimization, the total energies of native folds and incorrect protein models were approximately the same. The authors concluded that the energies obtained from standard force field calculations cannot be used to distinguish between correct and incorrect protein conformations. On the other hand, the investigation has shown that the packing criteria of the incorrect models were different from those normally found in native proteins. The incorrect structures clearly violated the general principles of close packing, hydrogen bonding, minimum exposed non-polar surface area, and solvent accessibility of charged groups. Examination of the interior showed that the packing of side chains at the secondary structure interfaces also differed from the characteristics observed in natural proteins (e.g. side chain ridges and grooves spirally wound on α-helices, predominantly flat surfaces of β-sheets). This analysis has clearly shown that the validity of model-built structures can only be assessed by a careful inspection of the structural features of a protein model.

For that reason several methods have been developed which try to distinguish between correct and incorrect folded protein structures [37–45]. One of these approaches is the 3D-Profiles method [37–40], which is based on the general principle that the 3D structure of a protein must be compatible with its own amino acid sequence. It measures this compatibility by reducing the 3D structure of the protein model to a simplified one-dimensional (1D) representation, the so-called *environment string*. The environment string has the same length as the corresponding amino acid sequence. This 1D string then can be compared with the respective amino acid sequence, which is also a 1D parameter.

In a first step the 3D structure of the protein model must be converted into a 1D parameter. For that reason the program determines several features of the environment of each residue: the area of the side chain that is buried in the protein; the fraction of the side chain area that is exposed to polar regions; and the secondary structure to which the particular amino acid belongs. Based on these characteristics each residue position is categorized into an environment class. A total of 18 distinct environment classes are implemented in the program [39]. In this manner the 3D structure is translated into a 1D string which represents the environment class of each residues in the protein model.

Although the environment string is 1D, it cannot be aligned with an amino acid sequence without some measure of compatibility for each of the distinct environment classes with each of the 20 naturally occurring amino acids. For that reason the program includes a compatibility scoring matrix (comparable with the scoring matrices described in section 4.3.1), which has been derived from sets of known protein structures [40]. Applying this compatibility matrix, the environment string and the amino acid sequence are aligned and a so-called 3D–1D score is obtained for the particular alignment. For obvious reasons it is more convenient to calculate local 3D–1D scores for small and medium-sized regions of about 5–30 residues length, than a global score for the complete alignment. The local scores are then plotted against residue positions to reveal local regions of relatively high or low compatibility between the 3D structure and the amino acid sequence [39]. Regions showing unusually low scores are likely to be regions where the protein conformation is incorrect, or where structural refinement is necessary.

The folding reliability can be also tested using knowledge-based force field methods [43–45]. These methods are based on the compilation of potentials of mean forces from a database of known 3D protein structures. The basic idea of these approaches is that atom–atom interactions in proteins are the primary determinant of proper protein folding.

A program named PROSA-II has been developed, which uses the mean force potentials to calculate the total energy of amino acid sequences in a number of different folds [43]. The calculated total energy of a particular protein conformation is a qualitative criterion for the confidence or quality of a predicted protein model. This is in contrast to investigations where the total energies, derived from standard molecular mechanics force fields, have been used to estimate the reliability of different protein conformations [36]. To test the predictivity of PROSA-II different native and incorrectly modified protein conformations have been used as a test set. It has been shown that for a very large number of proteins the derived total energy of the correctly folded protein is much lower than for any alternative (incorrect) protein conformation. Therefore, the program can be successfully applied to recognize erroneous protein folds or to detect faulty parts of structures in protein models.

References

[1] Bränden, C. J., and Jones, T. A. *Nature* **343**, 687–689 (1990).

[2] Jones, T. A., Zou, J. Y., and Cowan, S. W. *Acta. Cryst.* **A47**, 110–119 (1991).

[3] Engh, R. A., and Huber, R. *Acta. Cryst.* **A47**, 392–400 (1991).

[4] Hubbard, T. J. P., and Blundell, T. L. *Protein Eng.* **1**, 159–171 (1987).

[5] Drenth, J. *Principles of Protein X-ray Crystallography.* Springer Verlag: New York 1994.

[6] Morris, A. L., MacArthur, M. W., Hutchinson E. G., and Thornton, J. M. *Proteins Struct. Func. Gen.* **12**, 345–364 (1992).

[7] Laskowski, R. A., MacArthur, M. W., Moss, D. S., and Thornton, J. M. *J. Appl. Cryst.* **26**, 283–291 (1993).

[8] Hooft, R. W. W., Vriend, G., Sander, C. and Abola, E. E. *Nature* **381**, 272–272 (1996).

[9] Pontius, J., Richelle, J. and Wodak, S.J. *J. Mol. Biol.* **264**, 121–136 (1996).

[10] Ponder, J., and Richards, F. M. *J. Mol. Biol.* **193**, 775–791 (1987).

[11] Creighton, T. E. *Proteins: Structures and Molecular Properties.* 2nd Ed. W. H. Freeman and Company: New York 1993.

[12] Stewart, D. E., et al. *J. Mol Biol.* **214**, 253–260 (1990).

[13] Richards, F. M. *J. Mol. Biol.* **82**, 1–14 (1974).

[14] Richards, F. M. *Annu. Rev. Biophys. Bioeng.* **6**, 151–176 (1977).

[15] Chothia, C. *Annu. Rev. Biochem.* **53**, 537–572 (1984).

[16] Zehfus, M. H., and Rose, G. D. *Biochemistry* **25**, 5759–5765 (1986).

[17] Janin, J., and Chothia, C. *J. Mol. Biol.* **143**, 95–128 (1980).

[18] Leszczynski, J. F., and Rose, G. D. *Science* **234**, 849–855 (1986).

[19] Kabsch, W., and Sander, C. *Biopolymers* **22**, 2577–2637 (1983).

[20] Frishman, D., and Argos, P. *Proteins Struct. Func. Gen.* **23**, 556–579 (1995).

[21] Hooft, R.W., Sander, C. and Vriend, G. *Comput. Appl. Biosci.* **13**, 425–430 (1997).

[22] Hunt, N. G., Gregoret, L. M., and Cohen, F. E. *J. Mol. Biol.* **241**, 214–225 (1994).

[23] Laskowski, R. A., Thornton, J. M., Humblet, C. and Singh, J. *J. Mol. Biol.* **259**, 175–201 (1996).

[24] Privalov, P. L., and Gill, S. J. *Adv. Protein Chem.* **39**, 191–234 (1988).

[25] Chothia, C. *J. Mol. Biol.* **105**, 1–12 (1976).

[26] Wolfenden, R., Anderson, L., Cullis, P. M., and Southgate, C. B. *Biochemistry* **20**, 849 (1983).

[27] Miller, S., Janin, J., Lesk, A. M., and Chothia, C. *J. Mol. Biol.* **196**, 641–656 (1987).

[28] Lee, B., and Richards, F. M. *J. Mol. Biol.* **55**, 379–400 (1971).

[29] Eisenberg, D., and McLachlan, A. D. *Nature* **319**, 199–203 (1986).

[30] Lijnzaad, P., Berendsen, H. J. and Argos, P. *Proteins Struct. Func. Gen.* **25**, 389–397 (1996).

[31] Rose, G. D., Geselowitz, A. R., Lesser, G. L., Lee, R. H., and Zehfus, H. *Science* **229**, 834–838 (1985).

[32] Rose, G. D., and Dworkin, J. E. The Hydrophobicity Profile. In: *Prediction of Protein Structure and Function and the Principles of Protein Conformation.* Fasman, G. D. (Ed.). Plenum Press: New York; 625–634 (1989).

[33] Rashin, A., and Honig, B. *J. Mol. Biol.* **174**, 515–521 (1984).

[34] Rao, S. T., and Rosman, M. G. *J. Mol. Biol.* **76**, 214–228 (1973).

[35] Chothia, C., and Lesk, A. M. *EMBO J.* **5**, 823–826 (1986).

[36] Novotny J., Bruccoleri, R., and Karplus, M. *J. Mol. Biol.* **177**, 787–818 (1984).

[37] Fischer, D. and Eisenberg, D. *Curr. Opin. Struct. Biol.* **9**, 208–211 (1999).

[38] Bowie, J. U., Lüthy, R., and Eisenberg, D. *Science* **253**, 217–221 (1990).

[39] PROFILES-3D, User Guide, Accelrys, San Diego U.S.A.

[40] Lüthy, R., McLachlan, A. D., and Eisenberg, D. *Proteins Struct. Func. Gen.* **10**, 229–239 (1991).

[41] Novotny, J., Rashin, J. J., and Bruccoleri, R. E. *Proteins Struct. Funct. Gen.* **4**, 19–25 (1988).

[42] Hendlich, M., Lackner, P., Weitckus, S., Floeckner, H, Froschauer, R., Gottsbacher, K, Casari, G., and Sippl, M. J. *J. Mol. Biol.* **216**, 167–180 (1990).

[43] Domingues, F.S., Koppensteiner, W.A., Jaritz, M., Prlic, A., Weichenberger, C., Wiederstein, M., Floeckner, H., Lackner, P., Sippl, M.J. *Proteins Struct. Func. Gen.* **37**, 112–120 (1999).

[44] Casari, G., and Sippl, M. J. *J. Mol. Biol.* **224**, 725–732 (1992).

[45] MacArthur, M. W., Laskowski, R. A., and Thornton, J. M. *Curr. Opin. Struct. Biol.* **4**, 731–737 (1994).

4.6
Properties of Proteins

4.6.1
Electrostatic Potential

As we have already mentioned electrostatic interactions are among the most important factors in defining the conformation of a molecule in aqueous solution and in determining the energetics of interaction between two approaching molecules. The protein itself, the solvent, cofactors and prosthetic groups are nearly always charged or dipolar, and so the range of effects which are dependent in one way or another from electrostatics is broad [1–4]. Contrary to dispersion forces the electrostatic interactions are effective over relatively large distances. Due to their strong influence on structure and function of macromolecules in aqueous solution, it is absolutely necessary to consider explicitly the electrostatic term in any theoretical study on proteins [1]. For this purpose theoretical models are needed, which are able to describe correctly the electrostatic effects in proteins.

The interaction between any two charges is described by Coulomb's law (see section 2.2.1) which, in its simplest form is only valid for two point charges in vacuum. If the charges are immersed in any other matter, then particles of the surrounding matter are polarized by the presence of the charges, and the induced dipoles of the particles interact with the original point charges. Thus, the total resolved force on each of the point charges is altered, and the electrostatic interaction is decreased by the influence of the dielectric medium.

In classical electrostatic approaches, the materials are considered to be homogeneous dielectric media, which can be polarized by charges and dipoles. A dielectric constant is used as a macroscopic measure of the polarizability of a medium rather than explicitly accounting for the polarization of each atom. The portrayed procedure is called a *continuum model.*

It must be borne in mind that this view is simplistic and that the concept of dielectric constant—which constitutes a genuine macroscopic property—is valid only for homogeneous media. Less homogeneous environments must be treated explicitly. Special problems arise at the boundaries between regions of very different dielectric properties [5]. The surface of a protein represents such a case, because it divides the molecule into two regions which differ dramatically in composition. The molecular interior possesses a very low dielectric constant and includes a particular number of charges (most of them near the surface). Outside the protein there is a polar aqueous medium, which normally contains a distinct quantity of ions. For two point charges separated by a specific distance in a macromolecule in aqueous solution, the electrostatic interaction energy depends on the shape of the macromolecule and the exact positions of the charges (for a detailed description of this topic, see [5, 7, 8]. When using Coulomb's law for the calculation of electrostatic interactions, this fact will not be taken into consideration.

The multiple interactions occurring among the point charges and dipoles of the protein and the solvent are mutually dependent and turn the simple relationship of

Coulomb's law into a very complex state. The electrostatic interaction among molecules in a homogeneous environment can be averaged and expressed, as we have seen above, with the help of a simple dielectric constant. This concept is not valid for the inhomogeneous environment of proteins. Their electrostatic properties involve interactions among the multiple charges and dipoles of the proteins, and between these and the surrounding solvent and any ions that it contains. In this situation interactions between particular charges and dipoles must be calculated individually. This is impractical with the many atoms of the protein and solvent.

The major problem studying electrostatic effects in proteins is, as we have seen, the treatment of polarization effects [4]. In many electrostatic problems, real materials are treated as simple continua, and the effects of the underlying microscopic structure of the material is only incorporated into the macroscopic dielectric constant. At the microscopic level, the shielding of the charges arises from the polarizability of the individual atoms. Thus, an approach which discards the use of a dielectric constant and considers the individual atoms of the system and their mutual polarizabilities would be the best way to solve the problem. Of course, the exact quantum mechanical treatment would be a suitable solution, but this at present— due to limitations of computer power—is not practicable for systems of the size of proteins. Therefore, empirical approaches are generally employed for the exact calculation of electrostatic interactions within proteins [6–12].

Most of these approaches make use of the point charge approximation, i.e. the charge distribution of a protein is described by locating point charges at the atom centers. Several methods have been developed to obtain corresponding partial charges [13–15]. The procedures used are comparable with those described for the small molecules (see section 2.4.1.1). Because the complete protein is too large for a quantum mechanical charge calculation, the charges have been calculated for smaller fragments, like individual amino acids. The so-derived point charges for individual atoms of particular amino acids are then stored in point charge libraries from which they can be retrieved and assigned to each atom in the protein of interest. The often-used Kollman charges, for example, have been determined by scaling point charges to fit the ab initio-derived molecular electrostatic potential [14]. In the case of proteins, the ionization state has also to be taken into consideration. Therefore, formal charges are assigned to those amino acid residues that are expected to exist in charged state under physiological conditions. These charges are placed on one or two of the atoms of a residue. For example, an aspartic acid residue obtains the formal charge −1, which is assumed to be distributed over the two carboxylic oxygen atoms.

In one of the first approaches for a more reliable consideration of electrostatic interaction within proteins, the use of a distance-dependent dielectric constant was introduced. The mathematical equation used for the respective function often has the form $\varepsilon(r) = r$, where r is the distance between the atoms of interest [16]. The distance-dependent dielectric constant is based on plausibility rather than on any experimentally measurable effect. It is assumed that at distances of the order of atomic dimensions the dielectric constant between two charges is that of vacuum conditions, and that at much larger separations the dielectric constant of water $\varepsilon = 80$

holds true. For intermediate distances it is assumed that the dielectric varies with distance in an appropriate way. Distance-dependent dielectric constants can partially mimic the solvent-screening effects on electrostatic energies and are sufficient to stabilize macromolecules in molecular dynamics simulations. However, they cannot correctly describe properties like the electrostatic forces and the electrostatic potential.

A solution of the electrostatic problem may be provided by the use of the Poisson–Boltzmann equation. This equation belongs to the class of differential equations that are typical for the description of boundary phenomena. The Poisson–Boltzmann equation provides a rigorous approach for the calculation of the electrostatic effects of proteins, including the electrostatic potential. Several procedures have been developed which make use of the Poisson–Boltzmann equation. Two commercially available programs are DelPhi [17, 18] and UHBD [10, 19].

In the framework of the Poisson–Boltzman approach the macromolecular system is considered to consist of two separate dielectric regions. The solvent-accessible surface of the protein defines the boundary between these two regions. The interior of this surface is defined as the solute and the exterior is defined as the solvent. Water molecules located in the interior of the protein are usually treated as part of the solute rather than of the solvent. The protein is described in terms of its 3D structure with the location of point charges on the atom centers. A low dielectric constant is used for all points inside the solvent-accessible surface. Common values for this parameter range from 2 to 5. The Poisson–Boltzmann equation is also able to consider the electrostacic effects associated with ions embedded in the solvent. Thus the physiological conditions (0.145 mol l^{-1}) can be incorporated in the calculation.

Use of the Poisson–Boltzmann approach yields the total electrostatic potential of a charged molecule in a solvent according to the following simplified equation:

$$\phi_i^{tot} = \phi_i^{coul} + \phi_i^{self} + \phi_i^{cross} + \phi_i^{own}$$

The solvent molecule responds to the electrostatic field generated by each point charge in the molecule. This response, which consists of two electrostatic effects, the dipolar orientation and the electronic polarization, in turn sets up an electrostatic field at the positions of the original point charges, which is called the *reaction field* [20]. The magnitude of the reaction field is determined by the point charge, its distance from the molecular surface, the shape of the surface and the dielectric constants of molecule interior and solvent. The reaction field exerts a force on all point charges in the system, including the source charge itself. The total electrostatic potential ϕ_i^{tot} is the sum of the interaction of each point charge with its self-reaction field ϕ_i^{self}, the reaction field induced by other point charges ϕ_i^{cross}, the direct coulombic interaction with other point charges ϕ_i^{coul}, and the intrinsic electrostatic potential generated by each point charge ϕ_i^{own} (for a detailed description of this topic, see [5, 8, 21]).

The Poisson–Boltzmann equation is actually a reliable model for the electrostatic interaction in proteins, because it considers the effect of polarization as well as the ionic strength. Unfortunately, it is a very complex differential equation and can be solved analytically only for small regular systems. The alternative to the analytical solution is the use of numerical techniques to find an approximate solution even for large protein systems. For the numerical solution the programs use the so-called *finite difference method* (FDPB). Herein, the protein is mapped onto a 3D cubic grid. The calculated values for the charge density and the electrostatic potential are located on each point of the cubic grid. The numerical solution yields values which are accurate to within 5% in comparison with analytical solutions (which are available for small systems). The most critical regions—and thus the regions of largest errors—are usually those located near charged residues on the protein surface. Several procedures have been recently developed to avoid these errors [18].

The Poisson–Boltzmann method not only offers the possibility to calculate the electrostatic potential of a protein. Additionally, parameters such as the total electrostatic energy of the system, the solvation energy, and the reaction-field energy of

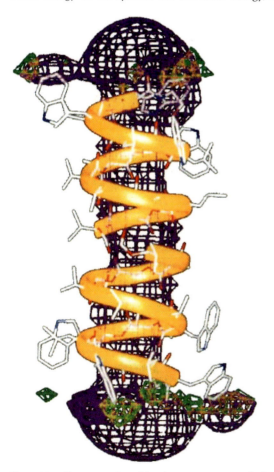

Figure 4.14 Representation of the electrostatic potential of a gramicidin A dimer embedded in a membrane environment. Calculations were performed using DelPhi. Color code: magenta = negative, green = positive potentials

proteins can be calculated. Nevertheless, the most important parameter is the electrostatic potential, which can be displayed in various ways (as described for small molecules in section 2.4.1.2).

Electrostatic potentials have been shown to play an important role in molecular recognition and binding. For example, the electrostatic potential of the superoxide dismutase enzyme has been shown to be responsible for enhanced external diffusion rates of the substrates to the active site [22]. The investigation of the electrostatic potentials of two trypsin enzymes, rat and cow trypsin, has yielded interesting results [23]. These two enzymes, although having the same catalytic mechanism, differ in net charge by 12.5 units. The calculation of the electrostatic potentials, using the Poisson–Boltzmann approach, revealed that both active sites are effectively shielded from the charges located on the surface, resulting in near-identical electrostatic potentials inside the active sites.

As one example for the graphical representation of the electrostatic potential of a protein, gramicidin A, a well-known membrane cation transporting protein, is shown in Fig. 4.14. Gramicidin A forms a dimer in the membrane. The calculation of the electrostatic potential has been performed for the gramicidin A dimer embedded in a low dielectric membrane layer (which is treated as part of the low dielectric solute system) using the program DelPhi.

4.6.2
Interaction Potentials

Other important features for studying interaction, recognition and binding of possible substrates to a protein are provided by the evaluation of molecular interaction fields. As we have already comprehensively discussed in section 3.2, interaction potentials are useful indicators for the prediction of binding properties of molecules. Programs, like the widely used GRID [24, 25], can be used to map regions within a protein where a water molecule or a substrate is attracted preferentially. The interaction fields, derived with a particular probe, can also be used as a starting point for docking studies of a substrate to its active site. The techniques and procedures applied in this context, are the same as those described in the case of small molecules in section 2.4.2. Various examples are given in literature where these programs have been used successfully to predict binding regions [26–28], to dock molecules into active sites [29–32] and to optimize structures of ligands in order to optimize the binding properties [26, 33, 34].

4.6.3
Hydrophobicity

As we have discussed in section 4.5.2 on the packing quality of proteins, the hydrophobic properties play an important role in the process of protein folding. Also, the protein binding reactivities are often determined by hydrophobic interactions. As was discussed for small molecules (section 2.4.3) several methods are available for the representation of hydrophobic and hydrophilic properties of molecules. The hydrophobicity can be either represented directly on the molecular surface or as a hydrophobic field in the space surrounding the molecule. Useful programs in this respect are, for example, GRID [24], HINT [35] and MOLCAD [36]. A detailed description of the different methods and a comparison of the results derived in studies on proteins is given in the literature [37].

References

[1] MacArthur, M. W., Laskowski, R. A., and Thornton, J. M. *Curr. Opin. Struct. Biol.* **4**, 731–737 (1994).

[2] Honig, B, and Hubbel, W. *Annu. Rev. Biophys. Biophys. Chem.* **15**, 163–193 (1986).

[3] Matthew, J. B. *Annu. Rev. Biophys. Biophys. Chem.* **14**, 387–417 (1985).

[4] Schutz, C. N. and Warshel, A. *Proteins Struct. Func. Gen.* **44**, 400–417 (2001).

[5] Warshel, A, and Aqvist, J. *Annu. Rev. Biophys. Biophys. Chem.* **20**, 267–298 (1990)

[6] Zauhar, R. J., and Morgan, R. S. *J. Comput. Chem.* **9**, 171–187 (1988).

[7] Gilson, M., Rashin, A., Fine, R., and Honig, B. *J. Mol. Biol.* **183**, 503–516 (1985).

[8] Harvey, S. C. *Proteins Struct. Func Gen.* **5**, 78–92 (1989).

[9] States, D. J., and Karplus, M. *J. Mol. Biol.* **197**, 122–130 (1987).

[10] Karplus, M. and McCammon, J. A. *Nat. Struct. Biol.* **9**, 646–652 (2002).

[11] Warwicker, J., and Watson, H. C. *J. Mol. Biol.* **157**, 671–679 (1982).

[12] Warshel, A., and Levitt, M. *J. Mol. Biol.* **103**, 227–249 (1976).

[13] Jorgensen, W. L., and Tirado-Rives, J. *J. Am. Chem. Soc.* **110**, 1657–1666 (1988).

[14] Weiner, P. K., and Kollman, P. A. *J. Comput. Chem.* **2**, 287–299 (1981).

[15] Abraham, R. J., Grant, G. H., Haworth, I. S., and Smith, P. E. *J. Comput.-Aided Mol. Design* **5**, 21–39 (1991).

[16] McCammon, J. A., Wolyness, P. G., and Karplus, M. *Biopolymers* **18**, 927–942 (1979).

[17] DelPhi User Guide, Biosym Technologies, San Diego, California, USA

[18] Gilson, M., Sharp, K., and Honig, B. *J. Comput. Chem.* **9**, 327–335 (1987).

[19] Antosiewicz, J., McCammon, J. A., and Gilson, M. K. *J. Mol. Biol.* **238**, 415–436 (1994).

[20] Bottcher, C. J. F. *Theory of Electric Polarization.* Elsevier Press: Amsterdam 1973.

[21] Gilson, M. K., McCammon, J. A., and Madura, J. D. *J. Comput. Chem.* **9**, 1081–1095 (1995).

[22] Sharp, K., Fine, R., and Honig, B. *Science* **236**, 1460–1463 (1987).

[23] Soman, K. Yang, A., Honig, B., and Fletterick, R. *Biochemistry* **28**, 9918–9926 (1989).

[24] Goodford, P. J. *J. Med. Chem.* **28**, 849–857 (1985).

[25] Wade, R. C., Clark, K. J., and Goodford, P. J. *J. Med. Chem.* **36**, 140–147 (1993).

[26] Reynolds, C. A., Wade, R. C., and Goodford, P. J. *J. Mol. Graphics* **7**, 103–108 (1989).

[27] Wade, R. C., and Goodford, P. J. *Br. J. Pharmacol. Proc. Suppl.* **95**, 588P (1988).

[28] Miranker, A., and Karplus, M. *Proteins Struct. Funct. Genet.* **11**, 29–34 (1991).

[29] Meng, E. C., Shoichet, B. K., and Kuntz, I. D. *J. Comput. Chem.* **13**, 505–524 (1992).

[30] Byberg, J. R., Jorgensen, F. S., Hansen, S., and Hough, E. *Proteins Struct. Funct. Genet.* **12**, 331–338 (1992).

[31] Stoddard, B. L., and Koshland, D. E. *Proc. Natl. Acad. Sci. U.S.A.* **90**, 1146–1153 (1993).

[32] Nero, T. L., Wong, M. G., Oliver, S. W., Iskander, M. N., and Andrews, P. R. *J. Mol. Graphics* **8**, 111–115 (1990).

[33] Varney, M. D., Marzoni, G. P., Palmer, C. L., Deal, J. G., Webber, S., Welsh, K. M., Bacquet, R. J., Bartlet, C. A., Morse, C. A., Booth, C. L., Herrmann, S. M., Howland, E. F., Ward, R. W., and White, J. *J. Med. Chem.* **35**, 663–676 (1992).

[34] Ocain, T. D., Deininger, D. D., Russo, R., Senko, N. A., Katz, A., Kitzen, J. M., Mitchell, R., Oshiro, G., Russo, A., Stupienski, R., and McCaully, R. J. *J. Med. Chem.* **35**, 823–832 (1992).

[35] Kellogg, G. E., Semus, S. F., and Abraham, D. J. *J. Comput.-Aided Mol. Design* **5**, 545–552 (1991).

[36] Heiden, W., Moeckel, G., and Brickmann, J. *J. Comput.-Aided Mol. Design* **7**, 503–514 (1993).

[37] Folkers, G., Merz, A., and Rognan, D. CoMFA as a tool for active site modelling. In: *Trends in QSAR and Molecular Modelling 92.* Wermuth, C. G. (Ed.). ESCOM Science Publishers B. V.: Leiden; 233–244 (1993).

5
Protein-based Virtual Screening

High-throughput screening (HTS) of chemical libraries is a well-established method for finding new lead compounds in drug discovery [1]. However, as the available databases become larger and larger, the costs of such screenings rise whereas the hit rates decrease [2]. One possibility to avoid these problems is not to screen the whole database experimentally but only a small subset, which should be enriched in those compounds likely to bind to the target. This preselection can be done by virtual screening (VS), a computational method to select the most promising compounds from an electronic database for experimental screening [3]. Virtual screening can be carried out by searching databases for molecules fitting either a known pharmacophore [4] or a three-dimensional (3D) structure of the macromolecular target [5]. This chapter is not aimed at reviewing all computational procedures to screen an electronic database, but rather will be focused on protein-based virtual screening. The four essential steps of any virtual screening process (preparation, docking, scoring, post-filtering; see Fig. 5.1) will be dealt with in the following sections.

5.1
Preparation

5.1.1
Database Preparation

Compound libraries used in lead finding programs should generally be filtered first to remove unsuitable compounds that would not reach and pass clinical trials anyway, due to undesired properties. There are two ways of removing "non-druglike" molecules from databases. The first possibility is to use a series of different filters, each one excluding compounds with certain properties. Highly reactive and toxic compounds can be removed according to reactive moieties such as acyl-halides, sulfonyl-halides, Michael acceptors etc. [6]. The probably best known method to evaluate drug-likeness is the Lipinski "Rule-of-Five" [7] suggesting that poor absorption or permeation are more likely when the molecular weight is over 500, the calculated octanol/water partition coefficient (clogP) is higher than 5, when there are more than 10 hydrogen bond acceptors and more than 5 hydrogen bonds donors. All com-

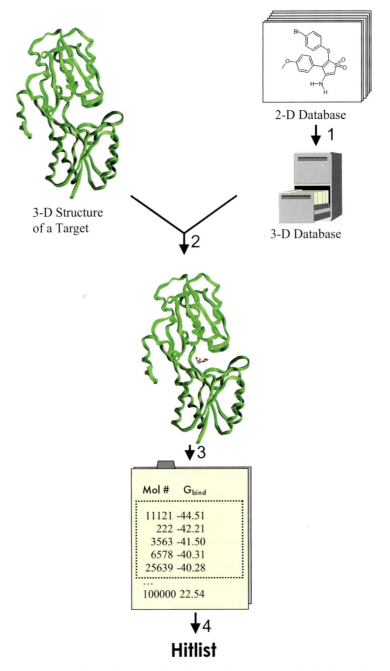

Hitlist

Fig. 5.1 Protein-based virtual screening flowchart. (1) Filtering out the 2-D database and convert 'drug-like' molecules in 3D, (2) docking each molecule of the chemical database in the active site of interest, (3) scoring the interaction of each ligand with the target protein, if a docking solution has been found, (4) extraction and post-processing of the top scorers to define a virtual hitlist.

pounds that fulfil two or more of these conditions are likely to show poor permeability and should thus be removed from the database. Meanwhile, more elaborated filters for specific absorption/distribution/metabolism/excretion (ADME) properties are being developed, such as filters for prediction of aqueous solubility [8], membrane permeation [9] and metabolic clearance [10]. A second possibility of filtering out unsuitable molecules is to design a universal filter from known databases that automatically distinguishes between drugs and chemicals. The filter then assigns to each compound either a "drug-likeness" score, or classifies the compounds simply as "drug" and "non-drug". Approaches have been published that are based either on neural networks [11], genetic algorithms [12], decision trees [13] or pharmacophoric description [14].

Apart from these "non-druglike" compounds, it might be desirable to eliminate molecules with a high propensity to bind to a wide array of protein targets. Such compounds are often called frequent hitters [15] as they can perturb the assay or detection method (e.g. fluorescent molecules) or non specifically bind to several macromolecular targets through aggregation or denaturation [20, 21]. In any case, the filtering level (high, low) should be adapted to the size of the electronic database and the properties that are required for selection (hit, lead, pharmacological reagent).

5.1.2
Representation of Proteins and Ligands

Reproducing the conformational space accessible to a macromolecule is a very difficult task and always necessitates an approximation. Docking procedures can thus be classified into three categories depending on the approximation level; (i) rigid body docking: both protein and ligand are treated as rigid bodies; (ii) semi-flexible docking: the ligand only is considered flexible; (iii) fully flexible docking: both ligand and protein are treated as flexible molecules.

5.1.2.1 **Protein Flexibility**
Proteins are highly flexible molecules that exist in a range of conformational substates separated by low-energy barriers. Protein motions can be classified into three groups [5]: (i) Small-scale fast motions that involve mainly side-chain movements, but also some movements of backbone atoms. Binding sites frequently display enhanced flexibility compared to other areas of the protein surface. (ii) Large-scale, slow domain motions are for example hinge-bending movements, where rigid domains are connected by flexible joints which tether the domains and constrain their movement. Hinge-bending is believed to allow the "induced fit" of the ligand. (iii) Renaturation upon ligand binding affects many proteins which are in a partially unfolded state due to either a small hydrophobic core or the presence of uncompensated buried charges. Ligand binding stabilizes the bound conformation, thereby shifting the equilibrium. Conformational changes of the protein upon ligand binding can thus reach from only few side-chain movements to large hinge-bending movements. Obviously, proteins should be treated as flexible in ligand docking,

especially when using a ligand-free structure as the target. However, for a long time, this has been computationally not feasible. The development of docking algorithms taking protein flexibility into account has started only recently, aided by the exponential growth of computer processor speed, working memory and disk capacity.

One approach to handle protein flexibility is to start from a single 3D structure and then allow defined substructures to move. Leach et al. [18, 19] use a method that explores the conformational space of protein side-chains using a rotamer library. Keeping the protein backbone rigid, the algorithm tries to find the global minimum energy combination of amino acid side-chains and ligand conformations [18, 19]. Sandak et al. [20] introduced a completely different approach that does not take side chain flexibility into account, but allows hinge-bending movements of domains, an important motion in induced fit. Large stable regions of the protein (domains) are identified and modeled as being connected by three-dimensional rotational hinges. While the domains are allowed to move with respect to each other in the docking process, the domain itself is held rigid [20]. Schnecke et al. [21] described a virtual screening tool (SLIDE) that models ligand and protein flexibility as a post optimization process by resolving collisions between the placed ligand and the protein by directed rotations of the ligand or the side chains of the protein. Their results show that, in spite of introducing protein flexibility, the choice of the structure used as target is still very important for the virtual screening outcome [21].

Another group of methods uses an ensemble of several possible conformations derived experimentally, i. e. by X-ray diffraction or NMR, or computed from molecular dynamics or Monte Carlo simulations. One of the first approaches employing conformational ensembles was described by Knegtel et al. [22] who used a composite grid ("energy-weighted average" or "geometry-weighted average") incorporating multiple crystal and NMR structures of protein-ligand complexes. This mean structure is then taken to calculate for any probe atom the interaction energies in the composite grid [22]. The grids are then used for molecular docking [22–24] by a simple readout of the interaction energy values previously stored.

Claussen et al. recently introduced the flexible docking tool FlexE [25] which uses a unified protein description from superimposed structures. Similar parts of the structures are merged whereas dissimilar areas are treated as separated alternatives [25]. The different structures of the ensemble can be combined to form new overall structures during the docking process. Moreover, side chain flexibility is introduced using a rotamer library. FlexE is derived from the popular FlexX algorithm [26] that will be described later (see section 5.2). All substantial concepts like the representation of intermolecular interactions, the incremental construction algorithm and the scoring functions are therefore taken from the original FlexX code and adapted to the ensemble approach [25].

Including protein flexibility in the docking process is especially important when using low-resolution computed models. Since the draft of the human genome [27] is likely to bring an increasing number of pharmaceutical targets for which only the primary sequence is known, there is an increasing interest in using such models as targets in drug design. Schafferhans et al. [28] have developed a special approach (DragHome) to dock ligands to approximate protein models which combines information from homology modeling with 3D QSAR ligand data. The binding site of a protein model is analyzed in terms of putative ligand interaction sites and translated via Gaussian functions into a functional binding site. Ligands to be docked onto these binding site representations are similarly translated into a description based on Gaussian functions. The docking is then computed by optimizing the overlap between the functional description of the binding site and the ligand [28]. Recently, another approach was introduced from Wojciechowski et al [29] in which both the ligand and the receptor are represented by approximate discretized models compensating the numerous structural inaccuracies resulting from the theoretical predictions of the receptor structure. Using an exhaustive scan over the six-dimensional translational and rotational degrees of freedom of the ligand, the best relative orientation of the two molecule is then computed by means of their sterical and chemical complementarity [29].

5.1.2.2 Ligand Flexibility

Since drugs are much smaller than macromolecules, ligand flexibility is computationally easier to handle and thus standard in today's docking routines.

The simplest approach is to store multiple conformations of the ligands in the database, each conformation being regarded as rigid during the docking process [30]. Subsequently, each conformation of a ligand is docked individually. Lorber et al. [31] used, however, a much faster method in which the conformers are docked together as an ensemble into the receptor binding site. For an ensemble of n ligand conformations, the orientation in the binding site is only calculated once, and the evaluation of the fits, which will differ from conformation to conformation, is performed n times [31].

Another way of introducing ligand flexibility is to store only one conformation per ligand in the database, but then to treat the ligands as flexible molecular entities during the docking process. A common strategy is the incremental construction method, which divides the ligand into fragments and incrementally builds it up in the receptor binding site. Upon placement of a fragment, only the most likely dihedral angles are searched for the best solutions. Examples for docking programs using this method are FlexX [26] and Dock [23]. Another possibility is to encode flexibility around dihedral angles as genes into genetic algorithms as in Gold [32] or AutoDock3.0 [33]. Genetic operations (crossing over, mutation) that specifically target these genes will address ligand flexibility *per se*. Finally, exhaustive computational techniques for searching conformational space (molecular dynamics, simulated annealing, Monte Carlo) can be used under the condition that the pace of the search algorithm does not compromise the entire VS process [34].

5.2
Docking Algorithms

Currently available docking tools differ in aspects such as the description of molecular interactions, the algorithms used to create ligand structures and the average run-time per molecule. The algorithms can be grouped into those with deterministic and those with stochastic approaches. Deterministic algorithms are reproducible, whereas stochastic algorithms include a random factor and are thus not fully reproducible.

5.2.1
Incremental Construction Methods

In an incremental construction algorithm the ligand is not docked completely at once, but is instead divided into single fragments and incrementally reconstructed inside the active site (Fig. 5.2).

The first incremental construction method was the program Dock [23]. The first step in Dock is the identification of points in the active site where ligand atoms may be located. These points, called sphere centres, are identified by generating a set of overlapping spheres, which fill the site. These sphere centres try to capture shape characteristics of the active site with a minimum of points. The ligand is partitioned along each flexible bond to generate rigid segments. An anchor fragment is then selected from the rigid fragments either manually or automatically. This anchor fragment is oriented within the active site, independently of the remaining parts of the ligand, by "matching" ligand atoms with sphere centres. Dock is, therefore, sometimes also called fast shape matching algorithm. All possible anchor placements are scored in terms of their interactions with the protein (for implemented Dock scoring function see section 5.3), and the best ones are used for subsequent "growing" of the ligand. Finally, the conformations of the complete ligand with the best scores are selected. Testing the efficacy and selectivity of Dock4.0 for virtual screening on the example of the thrombin and progesterone receptors, Knegtel et al. found that flexible docking with Dock4.0 is indeed able to effectively prioritize molecules in screening databases [35].

FlexX also treats the ligands as flexible and the protein as rigid. Similarly to Dock, it divides the ligand along its rotational bonds into rigid fragments, docks first a base fragment into the active site and then reattaches the remaining fragments [26]. It differs, however, significantly from Dock in the method used for determining the placement of the base fragment. Rather than defining points where ligand atoms may be located, FlexX defines interaction sites for each possible interacting group of the active site and the ligand [26]. The interaction sites are assigned an interaction type (hydrogen bond acceptor, hydrogen bond donor, etc.) and are modeled by an interaction geometry consisting of an interaction centre and a spherical surface The base fragment is oriented by searching for positions where three interactions between the protein and the ligand can occur [32]. The remaining ligand components are then incrementally attached to the core [32]. At each growing step, a list of preferred torsional angle values is read and the best conformation in terms of protein-ligand interactions is maintained for further "growing" of the ligand (for implemented FlexX scoring function see section 5.3) [26]. Finally, the conformations of the complete ligand with the highest score are selected. Various adaptations of

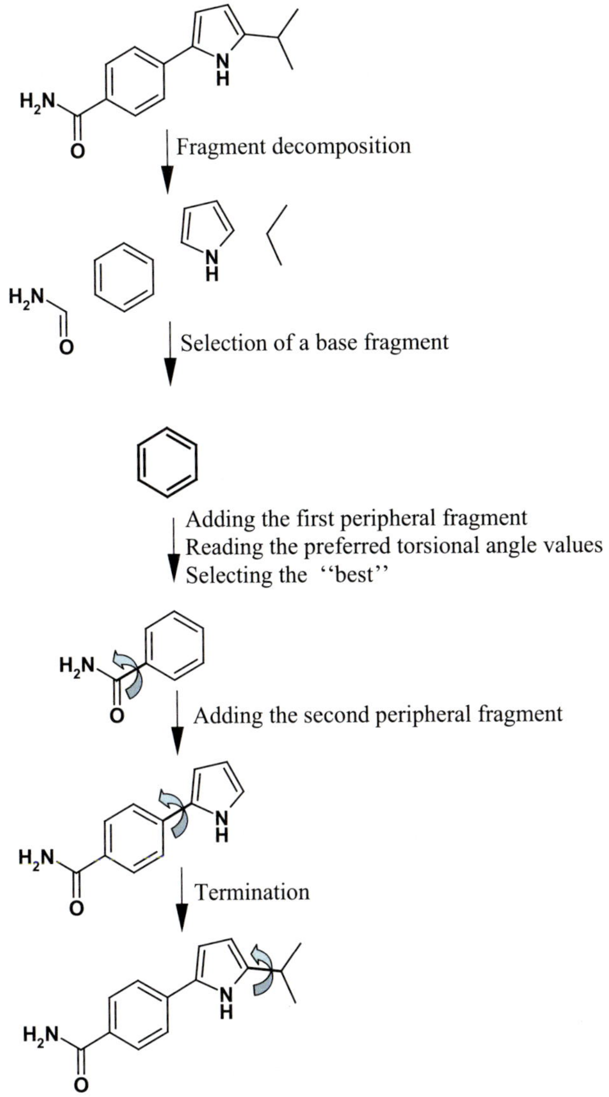

Fig. 5.2 Incremental construction of a ligand.

FlexX have recently been proposed for (i) flexible docking to a flexible target (FlexE, [25]), (ii) docking of combinatorial libraries (FlexC, [36]), (iii) pharmacophore-constrained docking (FlexX-Pharm, [37]).

Dock and FlexX are by far the most widely used docking tools using a fragment-based approach. Other programs such as Adam [38], Hammerhead [39] or Slide [21] are relatively similar to Dock and FlexX in generating base fragments but use alternative scoring functions (compare section 5.3) to prioritize placement of intermediate fragments.

5.2.2
Genetic Algorithms

A genetic algorithm (GA) is a computer program that mimics the process of evolution by manipulating a collection of data structures called chromosomes (Fig. 5.3). Each of these chromosomes encodes a possible solution to the problem to be solved. Gold [32] used such a GA for docking a ligand to a protein. Each chromosome encodes a possible protein-ligand complex conformation. A chromosome is assigned a fitness score based on the relative quality of that solution in terms of protein-ligand interactions. Starting from an initial, randomly generated "parent" population of chromosomes, the GA repeatedly applies two major genetic operators, crossover and mutation, resulting in "children" chromosomes that replace the least fit members of the population. The crossover operator requires two parents and produces two children, while the mutation operator requires one parent and produces one child. Crossover thus combines features from two different chromosomes into one, while mutation introduces random perturbations. The "parent" chromosomes are randomly selected from the existing population with a bias towards the best, thus introducing an evolutionary pressure into the algorithm. This emphasis on the survival of the fittest individuals ensures that, over time, the population should move towards an optimal solution, i.e. to the correct binding mode.

AutoDock 3.0 [33] uses a Lamarckian genetic algorithm (LGA). The characteristic of a LGA is that environmental adaptations of an individual's phenotype are transcribed into its genotype, based on Jean Baptiste de Lamarck's assertion that phenotypic characteristics acquired during an individual's lifetime can become heritable traits [40]. In AutoDock 3.0, each generation is thus followed by a local search (energy minimization) on an user-defined proportion of the population and resulting ligand coordinates are stored in the chromosome, replacing the parent [33].

1. **A set of operators** (crossing over, mutation, etc...) is chosen. Each operator is assigned a **weight**.
2. **An initial population** is randomly created and the **fitness** of all individuals computed.
3. **An operator is chosen** using a roulette wheel selection based on operator weights.
4. **The parents able to reproduce** are chosen using roulette wheel selection based on fitness scores.
5. **The genetic operator is applied**, children chromosomes generated, and the fitness of each child determined.
6. If not already present in the population, **children replace the least-fit individuals**.
7. **Goto 3** unless the maximum number of genetic operations is reached.

Fig. 5.3   Main features of a genetic algorithm.

GA algorithms are now commonplace for searching conformational space and several GA implementations have been recently described in combination with local minimisation strategies using various force fields [41–43].

5.2.3
Tabu Search

Tabu search (TS) algorithms have been first implemented in the Pro_Leads docking software [44]. A TS is characterized by imposing restrictions to enable a search process to negotiate otherwise difficult regions (Fig. 5.4). These restrictions take the form of a tabu list that stores a number of previously visited solutions. By preventing the search from revisiting these regions, the exploration of new search space is encouraged. Only one current solution is maintained during the course of a search. At the start of a run, the current solution is initialized by randomizing the position and orientation of the ligand within a certain box around the active site. From the current solution, a user-defined number of moves is generated by a mutation-like procedure and finally ranked according to a scoring function. The TS maintains a tabu list that stores a number of previously visited solutions, and a move is "tabu" if it generates a solution that does not differ sufficiently (e.g. RMS deviation < 0.75 Å) from the stored solutions. The highest ranked move is always accepted as new "current solution" if its energy is lower than the lowest energy obtained so far and replaces at the same time the previous "best solution". Otherwise, the algorithm chooses the best non-tabu move. If neither criterion can be met, the algorithm ter-

1. **Create an initial solution** at random. Make it the current solution
2. **Evaluate the current solution.** If best so far, record it as best solution
3. **Update the tabu list**
   (a) if tabu list not full, add current solution to the list.
   (b) Else, replace the oldest member with the current solution.
4. **Generate** and evaluate **x possible moves** (e.g. 1000) from the current solution.
5. Rank x moves in ascending order of interaction energy
6. **Examine the moves** in rank order.
   (a) if move has the lowest energy than the best solution so far, accept it and goto 7. (b) If move is not tabu, accept is and goto7.
   (c) If no acceptable move, **exit.**
7. If the **maximum number of iterations** is reached, **exit** with the best solution found.
   (a) If **best solution has not changed** for a number of iterations (e.g. 100) goto 1.
   (b) Else, goto 2.

Fig. 5.4 Main features of a tabu search.

minates. If a new current solution can be found, it is added to the tabu list. The new current solution is simply added to the end of the list until it is full (e.g. 25 solutions). Thereafter, the current solution replaces an existing solution stored in the tabu list in a "first-in, first-out" manner, that means it replaces the tabu solution having the longest residence in the list [42]. Once the new current solution has been identified, a new set of moves is generated from it and the search procedure continues with the next iteration [44]. TS and GA algorithms have recently been combined in SFDock [45] for limiting the conformational search space while docking a small molecule to a protein. While a GA usually converges quickly at the close proximity of a global minimum, it can become trapped in local minima. Using a tabu list helps in avoiding this drawback.

5.2.4
Simulated Annealing and Monte Carlo Simulations

Simulated annealing is a special molecular dynamics simulation in which the system is cooled down at regular time intervals by decreasing the simulation temperature. The system is thus trapped in the nearest local minimum conformation. Disadvantages of simulated annealing are that the result depends on the initial placement of the ligand and that the algorithm does not explore the solution space exhaustively. In a Monte Carlo search, the conformational space is sampled by random movements.

AutoDock2.4 uses Monte Carlo simulated annealing (MCSA) [46]. In MCSA, during each constant temperature cycle random changes are made to the ligand's current orientation and conformation. The new state is immediately accepted if its energy is lower than the energy of the preceding state. Otherwise, the configuration is accepted or rejected based upon a probability expression (Boltzmann equation). The probability of acceptance P is given as:

$$P = e^{\left(-\frac{\Delta E}{kT}\right)}$$

(1)

where ΔE is the difference in energy from the previous step, T is the absolute temperature in Kelvin, and k is the Boltzmann constant. This means, the higher the temperature of the cycle, the higher the probability that the new state is accepted [46].

In MCDock [47], a Monte Carlo simulation is used in two steps. The first step results in a purely geometrical optimization of the ligand position by random moves to minimize intermolecular overlap. It is followed by a Monte Carlo simulated annealing using the CHARMM force field. Both ProDock [48] and ICM [49] use internal coordinates to represent molecular structures and couple MC random moves with force field-based energy minimisation. DockVision [50], a very user-friendly docking tool, operates in three steps (i) random positioning of precalculated ligand conformations in the user-defined binding cavity, (ii) check for steric clashes with the target using a floating algorithm combining a MC procedure with a grid-based steric score function, (iii) search for optimal binding modes after relieving potential clashes by a force field-based MC sampling. QXP [51] combines random MC moves on dihedral angles of the ligand with energy minimisation with an intermediate fast template fitting procedure aimed at optimally locating the ligand in the binding cavity. Affinity [52] uses a two-step procedure to dock a flexible ligand to a partially flexible protein; (i) a classical MC procedure is applied to locate the ligand in the binding site, (ii) location is afterwards optimized by a simulated annealing protocol using a grid-based force field to treat the bulk (non movable part) of the complex and a more sophisticated full force field [53] including implicit solvation effects [54] is applied to the movable part of the system (ligand, amino acids of the binding site). Glide [55], another commercially available software uses a suite of hierarchical filters to remove unlikely solutions starting from low-level approximations (distance matches) to high-level calculations (full force field-based MCSA minimization) with free energy scoring. Glide implements a novel algorithm for rapid conformational generation minimizing computational costs by clustering the core regions of the generated 3D ligand conformations and treating the positions of the rotamer groups at the ends essentially independently.

5.2.5
Shape-fitting Methods

Shape-fitting methods are fast docking routines in which the steric and electrostatic complementarity of pre-calculated ligand conformations to the protein target is estimated. FTDock [56] uses a grid-based representation of both the ligand and the target with assignment of defined values on the grid that depend on the atom accessibility. Fast Fourier transforms are then used to optimize ligand-target complementarity after global rotation/translation of the ligand. Ligin [57] optimizes ligand orientation through a complementarity function summing up atomic contact surfaces to which a special weight is assigned depending on the favoured/disfavoured nature of the interaction. Sandock [58] takes into account both the steric and electrostatic complementarity of the ligand, fitted by a distance-matching algorithm to the protein accessible surface. Last, a fast rigid body docking procedure was recently described by Miller et al. [59] that uses a grid representation of the binding site for matching a triplet of ligand atoms to a triplet of hot spots (polar/apolar interaction sites). After elimination of improper orientation, possible matches are optimized by a soft atom pairwise potential.

5.2.6
Miscellaneous Approaches

One possible drawback of database docking may be the low diversity of selected hits. As a matter of fact, discriminating a true hit from structurally similar but inactive compounds is very difficult [60]. One possibility to overcome this problem is the "Docking by families" approach [61]. In this method, the database used for screening is grouped into families of related structures. All members of every family are docked using Dock, but only the best scoring molecule of a high-ranking family is allowed in the hit list. The identity and scores of the other members of these families are recorded as annotations to the best family members. A comparison of this "family-based docking" method with "molecule-by-molecule docking" showed that this method increased the diversity of the hit list and that more families of known ligands were found [61]. Lamb et al. [62] propose a modified version of Dock allowing to dock combinatorial libraries to a set of related targets by docking first a scaffold, then selecting the best substituents and finally comparing proposed hits within or between libraries [62]. Interestingly, the method was able to reproduce binding mode obtained from X-ray data and experimental binding data for protein mutants. Docking a ligand to a library of proteins is, however, still in its infancy. The only study of 'reverse docking' reported [63] uses Dock and a collection of 2700 protein cavities from the Protein Data Bank. 50% of predicted targets for two known drugs (4H-tamoxifen, vitamin E) could be confirmed experimentally. The long computing time required to screen the cavity database (10 to 20 days) for a single ligand is, however, a clear drawback of the method. Paul et al. [64] have set-up a database of 2150 binding sites from the PDB for inverse screening using a customized version of Gold [32]. One advantage of the method is that protein flexibility is taken into account *per se*, as multiple copies of the same enzyme co-crystallized with different inhibitors are stored in the database. When applied to the 'in silico' target assignment of biotin, streptavidin was ranked first among the most likely targets with a hit rate of 70% among the top 50 scorers.

5.3
Scoring Functions

The free energy of binding is given by the Gibbs-Helmholtz equation:

$$\Delta G = \Delta H - T\Delta S \qquad (2)$$

with ΔG giving the free energy of binding, ΔH the enthalpy, T the temperature in Kelvin and ΔS the entropy. ΔG is related to the binding constant K_i by the equation

$$\Delta G = -RT \ln K_i \qquad (3)$$

with R corresponding to the gas constant.

There is a wide variety of different techniques available for predicting the free energy of binding of a small molecule ligand based on the given 3D structure of a protein-ligand complex. These techniques differ significantly in accuracy and speed. If one wants to predict the difference in the free energy of binding between a ligand and a reference molecule, very accurate but time-consuming techniques such as free energy perturbation can be used [65]. If the aim is, however, to compare the free energies of hundreds or thousands of protein-ligand complexes as generated by virtual screening, much faster albeit less accurate scoring functions have to be used

[66]. Scoring functions can be grouped into (i) empirical scoring functions like Ludi [67–68], FlexX [26], ChemScore [69], Fresno [70], Score [71, 72], PLP [73], (ii) force field based functions like Dock [23] and (iii) knowledge-based potential of mean force functions like PMF [74], DrugScore [75], BLEEP [76], SMOG [77]. In a virtual screening, scoring functions are used for two purposes: During the docking process, they serve as fitness function in the optimization of the placement of the ligand. When the docking is completed, the scoring function is used to rank each ligand in the database for which a docking solution has been found. In principle, different scoring functions could be used for these two purposes, although the same function is usually utilized for both in most docking tools.

5.3.1
Empirical Scoring Functions

Empirical scoring functions use several terms describing properties known to be important in drug binding to construct a master equation for the prediction of binding affinity. Multilinear regression is used to optimize the coefficients to weight the computed terms, using a training set of protein-ligand complexes for which both the binding affinity and an experimentally determined high-resolution 3D structure is known. These terms generally describe polar interactions such as hydrogen bonds and ionic interactions, apolar interactions such as lipophilic and aromatic interactions, loss of ligand flexibility (entropy) and eventually also desolvation effects.

A breakthrough in deriving empirical scoring functions came with the introduction of Böhm's function developed for the *de novo* design program LUDI [67, 68], which is now implemented in a modified form in FlexX [26] and represents a typical empirical function:

$$
\begin{aligned}
\Delta G = \Delta G_0 &+ \Delta G_{rot} * N_{rot} \\
&+ \Delta G_{hb} \, \Sigma \, f(\Delta R, \Delta \alpha) + \Delta G_{io} \, \Sigma \, f(\Delta R, \Delta \alpha) \\
&+ \Delta G_{aro} \, f(\Delta R, \Delta \alpha) + \Delta G_{lipo} \, f^*(\Delta R)
\end{aligned} \tag{4}
$$

The ΔG coefficients are unknown and are determined by multilinear regression in order to fit the experimentally measured binding affinities. The first terms are a constant term and a term taking into account the loss of entropy during ligand binding by burial of rotatable bonds (ΔG_{rot}: energy loss per rotatable bond, N_{rot}: number of rotatable ligand bonds) [60]. ΔG_{hb} and ΔG_{io} give the binding energy for each optimal hydrogen bond and salt bridge, respectively. $f(\Delta R, \Delta \alpha)$ is a scaling function penalizing deviations from the ideal interaction geometry in terms of distance (ΔR) and angle ($\Delta \alpha$). An equivalent scaling function is used for the aromatic interactions (ΔG_{aro}). A lipophilic term (ΔG_{lipo}) is calculated as a sum over all pairwise atom-atom contacts. The function $f^*(\Delta R)$ accounts for contacts with a more or less ideal distance and penalizes forbiddenly close contacts.

One major disadvantage of the empirical scoring functions is the need of a training set to derive the weight factors of the individual energy terms. One can therefore expect an empirical scoring function to perform well only for proteins (e. g. metalloenzymes, proteases) similar to that used in the training set [66].

5.3.2
Force Field-based Scoring Functions

Force field-based scoring functions such as the Dock energy score [23] are based on the non-bonded terms of a classical molecular mechanics force field (e.g. AMBER, CHARMM, etc...). A Lennard-Jones potential describes van der Waals interactions, whereas the Coulomb energy describes the electrostatic components of the interactions. The non-bonded interaction energy takes the following form:

$$E = \sum_{i=1}^{lig} \sum_{j=1}^{rec} \left[\frac{A_{ij}}{r^{12}} - \frac{B_{ij}}{r^6} + 332 \frac{q_i q_j}{D r_{ij}} \right] \tag{5}$$

where A_{ij} and B_{ij} are van der Waals repulsion and attraction parameters between two atoms i and j at a distance r_{ij} , q_i and q_j are the point charges on atoms i and j, D is the dielectric function, and 332 is a factor that converts the electrostatic energy into kilocalories per mole. The main drawback of force field calculations is the omission of the entropic component of the free energy of binding. Therefore, care should be taken not to overestimate the larger and most polar molecules that usually yield the highest enthalpy interaction scores.

5.3.3
Knowledge-based Scoring Functions

A major disadvantage of empirical scoring functions lies in the fact that it is unclear to which extent they can be applied to protein-ligand complexes that were not represented in the training set used in deriving the master equation. Furthermore, empirical scoring functions dissect the protein-ligand binding free energy into all its physically meaningful contributions and try to evaluate them explicitly. Desolvation and entropy terms are, however, especially difficult to quantify.

A more recently developed approach avoiding these disadvantages uses knowledge-based scoring functions [74–77] with potentials of mean force. A potential of mean force converts structural information gathered from protein-ligand X-ray coordinates into Helmholtz free interaction energies of protein-ligand atom pairs. It is assumed that the more often a protein atoms of type i and ligand atom of type j are found in a certain distance r_{ij}, the more favourable is this interaction. A protein-ligand interaction free energy A(r) is then assigned to each interaction type between a protein atom of type i and a ligand atom type j in a distance r_{ij}, depending on its frequency.

$$A(r) = -k_B T \ln g_{ij}(r) \tag{6}$$

where k_B is the Boltzmann constant, T the absolute temperature and $g_{ij}(r)$ the atom pair distribution function for a protein-ligand atom pair ij. The distribution function is calculated from the number of occurrences of that pair ij at a certain distance r in a database of protein-ligand complexes (usually the PDB).

The score is defined as the sum over all interatomic interactions of the protein-ligand complex. Advantages of this approach are that no fitting to experimentally measured free energies of binding of the complexes in the training set is needed, and that solvation and entropic terms are implicitly included.

5.4
Postfiltering VS results

Docking and scoring a large database (>100 000 molecules) for complementarity to a protein active site usually yields several hundreds of virtual hits. There is thus a need to reduce the initial hitlist without losing any information about potential ligands. Post-filtering the hitlist should furthermore discard most of the false positives from the simple knowledge of either the ligand structure or the predicted 3D structure of the target-hit complex.

5.4.1
Filtering by Topological Properties

In order to eliminate protein-ligand complexes with improper geometries, different filters can be applied that evaluate three-dimensional protein-ligand complexes in terms of their steric complementarity. Stahl et al. [78] developed a set of such filters including the fraction of the ligand volume buried inside the binding pocket, the size of lipophilic cavities along the protein-ligand interface, the solvent-accessible surface of nonpolar parts of the ligand, and the number of close contacts between non-hydrogen-bonded polar atoms of the ligand and the protein. In an optimal conformation, the buried ligand volume should be as large as possible, whereas the size of lipophilic cavities, the solvent-accessible surface of nonpolar parts of the ligand and the number of close contacts between non-hydrogen-bonded polar atoms should be as small as possible.

5.4.2
Filtering by Multiple Scoring

The main reason for the presence of false positives in a virtual hitlist lies in the limited accuracy (7–10 kJ/mol) of empirical scoring functions [66]. As every scoring function has potential advantages and drawbacks, a logical scoring strategy is to use several scoring functions in combination to select only hits that achieve high scores from two or three different docking functions. This method named consensus scoring was introduced by Charifson et al. [79] and further used by Bissantz et al. [80]. Both studies showed that consensus lists generated from two or three different scoring functions contain significantly lower numbers of false positives than any hitlist obtained by a single scoring function, and conclude that an optimal combination of scoring functions significantly enhances hit rates [79–80]. As an alternative to consensus scoring, Stahl et al. [81] propose to focus hit selection on the very top scorers given by the best possible scoring function. Consensus scoring can be performed in several ways; (i) the "rank-by-number" approach ranks hits according to the average predicted values (in a similar scoring unit) given by different scoring functions; (ii) the rank-by-rank approach uses average ranks instead; (iii) the "rank-by-vote" protocol gives a vote from a particular function if a compound is predicted to be scored among the top N% scorers and sums up the number of votes gathered from all scoring functions. In a recent theoretical exercise, Wang et al. [82] recently concluded that consensus scoring outperforms any single scoring for a simple statistical reason: the mean value of repeated samplings tends to be closer to the true value. Both the rank-by-number and the rank-by rank strategy were shown to work more effectively than the rank-by-vote strategy [82]. Finally, Clark et al. [83] recently showed that consensus scoring reduces not only errors in properly ordering ligands with affinities in the nM to μM range, but also allows the identification of a good binding orientation among a set of possible solutions. A slightly different approach ("Multi-Score") introduced by Terp et al. [84] is aimed at improving compound selection by quantifying the protein-ligand affinities. 120 ligands for which both the three-dimensional structures and the binding affinities were known have been docked to their respective target and scored using eight different scoring functions. The scores obtained have then been fitted to the experimentally determined binding affinities using partial least-squares projections onto latent structures on the eight computed scores [84]. Multiscore scoring was shown to correlate rather well ($Q^2 = 0.60$) with experimentally-determined binding affinities whereas none of the eight individual scoring function did.

5.4.3
Filtering by Combining Computational Procedures

The methods described above are generally too time-consuming to be used in virtual screening of large database but can be readily applied to known ligands in a lead optimization research program. Several groups suggested approaches to improve the selection of the correct bioactive conformation out of the set of possible solutions. Hoffmann et al. [85] have introduced a two-stage method in an attempt to rerank the inhibitor conformations generated by the docking program. In the first step, a large number of ligand conformations is generated using FlexX. In the second step, these conformations are minimized using a classical force field and consequently reranked [85]. Another multistep procedure ('divide and conquer') was developed by Wang et al. [86]. After generating low-energy conformers of the ligand using a grid-based potential, the most likely conformations are rigidly docked to the binding site using Dock [23]. The resulting protein-ligand complexes are then refined by molecular mechanics minimization, conformational scanning at the binding site and finally a short simulated annealing [86]. Coupling docking analysis to structure-activity relationships (SAR) in a series of congeneric ligands has been proposed in the DoMCoSAR approach [87]. After docking a series of related ligands, the common binding mode of the largest substructure is taken as constraint to redock the compounds and the results are analyzed by computing interaction-based descriptors including solvation contributions. The docking mode that is most consistent with existing SAR is finally selected.

Terp et al. [84] developed yet another method (MultiSelect) that circumvents the second docking step and allows the selection of a single ligand conformation. After generating a set of possible docked ligand conformations using Gold [32], all conformations so generated are rescored using eight scoring functions. The obtained scores are consequently analyzed using principal component analysis to yield a single conformation. Recently, Paul et al. [88] introduced a new consensus docking approach (ConsDock) that uses three different docking tools (Dock, FlexX and Gold) for a consensus analysis of all docking conformations generated in four steps [88]: (i) hierarchical clustering of all conformations generated by a docking tool into families represented by a lead; (ii) definition of all consensus pairs from leads generated by different docking tools; (iii) clustering of consensus pairs into classes, represented by a mean structure; and (iv) ranking the different mean structures starting from the most densely populated class of consensus pairs. ConsDock was shown to outperform single docking in placing 60% of top-ranked conformations within 2 Å rmsd of the true structure as obtained by X-ray analysis.

5.4.4
Filtering by Chemical Diversity

Clustering by chemical similarity or diversity is probably the easiest way to reduce a virtual hitlist to a list of representative compounds. Many approaches can be followed using either fingerprints, interatomic distance pairs or scaffolds [89]. The concept of chemical diversity is described in detail in section 2.4.2.

5.4.5
Filtering by Visual Inspection

One cannot avoid looking at the predicted 3D structure of the protein-hit complex at the earliest possible stage to ascertain that (i) the ligand has been really docked in the expected binding site and not at its periphery, (ii) the bound conformation of the ligand has a physicochemical meaning, (iii) the ligand interacts with key residues of the active site. This step can be very time-consuming and tedious but ensures that every hit selected shows the expected properties.

5.5
Comparison of Different Docking and Scoring Methods

Independent comparisons of docking programs are still very rare, as it is extremely difficult to compare various algorithms having totally different settings and levels of approximation. In an independent challenge (CASP2) to predict the structures of 7 protein-ligand complexes that had been solved by X-ray diffraction but for which only the protein coordinates were revealed to the contenders, no docking method was found to perform significantly better than the others [90]. In approx. 50% of the cases, a solution within 2 Å rmsd from the conformation as obtained by X-ray could be found, but this was not necessarily the highest-ranked one. Using fast screening settings (i. e. with a pace of 90–120 s/ligand), Paul et al. [88] recently showed that Gold significantly outperforms FlexX and Dock in reproducing the X-ray structure of 100 protein-ligand complexes from the PDB, whatever the solution considered (top-ranked, closest to the true structure).

A comparison of various scoring functions is easier to perform as one can rely on high-resolution X-ray structures of protein-ligand complexes and simply score the interactions. Depending on the physicochemical properties of the binding cavity, the performance of a scoring function may vary considerably. For example, the FlexX scoring function [26] that contains a directional H-bonding term was shown to perform well for protein-ligand complexes forming several H-bonds and salt bridges but performed poorly when pure hydrophobic interactions had to be reproduced [91]. In contrast, the PLP scoring function is more sensitive to the overall steric fit [91]. Depending on the discovery stage (early hit identification, final lead optimization) different levels of approximation have to be taken into account. When applied to a small set of ligands in a late optimization phase, more sophisticated techniques (free energy perturbation, free energy grid methods) can be used as they clearly outperform empirical scoring functions [92].

For a couple of years, several groups have been trying to combine different docking methods with various scoring functions [79–81, 92–95]. All studies agree to conclude that the best combination is unpredictable and depends on molecular properties of the target. In all cases, consensus scoring provided hit lists enriched in true ligands by decreasing the number of false positives. A possible strategy proposed by Bissantz et al. [80] consists in a two-stage screening; (i) screening a small dataset seeded with a couple of known ligands to derive the best docking/scoring combination (implying that several docking tools have to be combined with several scoring functions), (ii) applying the optimized strategy to the screening of the full library. This knowledge-based virtual screening approach is, however, not applicable to orphan targets for which no ligand is known.

5.6
Examples of Successful Virtual Screening Studies

There are numerous examples of successful virtual screening (VS) applications in the literature (Table 5.1), most of them using the pioneering Dock program [23]. After the early rigid docking attempts [96–99, 101] that nevertheless yielded experimentally validated hits, flexible docking has now become commonplace. Most of the investigated targets are enzymes for which a high-resolution X-ray structure exists. There are only very few VS examples for receptors [107, 112]. In many cases [96, 98–99, 100–103, 105, 108, 110, 113, 116] validated hits share (i) a rather high molecular

Tab. 5.1 Examples of successful virtual screening applications

Target	Size	Database	Method	Hit Rate, %	Reference
cercarial elastase	8,800	ACD[a]	Dock	3.8	96
CD4 D1 domain	150,000	ACD	Dock	9.7	97
HGXPRTase	unknown	ACD	Dock	11,1	98
Thymidilate synthase	unknown	ACD	Dock	20.0	99
FK506-binding protein	unknown	ACD/CSD[b]	Sandock	unknown	100
HIV-1 gp41	20,000	ACD	Dock	12.5	101
Farnesyltransferase	219,390	ACD	Eudoc	19.0	102
Kinesin	110,000	ACD	Dock	13.8	103
HIV1 Tar RNA	153,000	ACD	Dock + ICM	25.0	104
Protein tyrosine phosphatase 1B	150,000	ACD	Dock	28.0	105
DNA Gyrase	350,000	ACD + RCI[c]	Ludi	5.0	106
Retinoic Acid receptor	153,000	ACD	ICM	6.2	107
HGXPRTase	599	Com. Lib.[d]	Dock	31.2	108
Guanine phosphoribosyl transferase	1,357	Comb.Lib	Dock	35.7	109
Bcl-2	207,000	NCI[e]	Dock	20.0	110
Human carbonic anhydrase II	90,000	Maybridge[f]/ LeadQuest[g]	FlexX	69.2	111
Retinoic acid receptor	153,000	ACD	ICM	6.6	112
Triosephosphatase isomerase	108,000	NCI	Dock	20.0	113
Estrogen receptor	1,100,000	ACD-SC[h]	Pro_Leads	56.7	114
Glyceraldehyde phosphate deshydrogenase	1,860	Comb. Lib.	Dock	16,6	115
Protein tyrosine phosphatase1B	235,000	ACD/ Biospecs[i] + Maybridge	Dock	34.8	116
AMPC lactamase	229,810	ACD	Dock	5.3	117

a MDL Available Chemicals Directory (118)
b Cambridge structural database (119)
c Roche compound inventory
d Combinatorial virtual library
e National Cancer Institute (120)
f Maybridge (121)
g TRIPOS LeadQuest (122)
h ACD screening collection (123)
i Biospecs (124).

weight, (ii) an experimentally-determined affinity in the high μM range (> 50 μM). It is questionable whether such compounds can really be classified as hits, since pharmaceutical companies would remove most of these compounds from HTS hit lists. They are likely to belong to the family of molecules (dyes, polyaromatic compounds, etc.) that have been recently shown to be either promiscuous binders [16] or non-specific ligands [17]. One major reason for this observation is that, until very recently, screened databases have not been extensively filtered to remove unsuitable molecules. An efficient automated procedure [125–127] to select 'drug-like' molecules from large datasets is thus a necessary preliminary step.

How to compare HTS with VS? An interesting study [116] recently addressed this question by comparing hits coming from either HTS or VS for the same target. Although different databases were compared, VS led to a 1700-fold higher enrichment rate in true hits than HTS. Interestingly, VS hits were found to be more drug-like than the HTS compounds and chemically different, suggesting that VS could ideally complement HTS. A good example of the power of VS has recently been given by Böhm et al. [106] who reported, for an enzyme that resisted to classical HTS, the successful identification of micromolar inhibitors by computer screening.

5.7
The Future of Virtual Screening

Which type of applications can be envisaged in the near future for VS ? At least four directions seem to already have shown some promise:

(i)   To identify non peptide ligands of targets for which peptidic ligands already exist [104, 105]. VS can be considered here as a logical follow-up in early discovery stages where peptide libraries provide the very first hits and pharmacophoric constraints.

(ii) To optimize existing ligands discovered by either random screening or in a prior VS stage [97, 98, 106, 108, 109]. This is not an easy exercise due to the still limited capacity of fast scoring functions to accurately predict the absolute free energies of binding. However, it may help the bench chemist to prioritize compounds for combinatorial/parallel synthesis.

(iii) To discover the very first ligands of orphan targets. The future of computer screening is likely to reside in this application, provided that sufficient structural information is known about the target. The genomic revolution [27] is already providing computer chemists with interesting pharmacological targets for which no ligands are known but a biological function could have been identified.

(iv) To discover ligands of targets for which precise 3D information (NMR, X-ray) is missing. Very few homology models [96, 107, 112] have been challenged for their capacity to identify novel ligands by computer screening. A promising study by Bissantz et al. [128] on the very important family of G-Protein coupled receptors (GPCR) suggests that it is indeed feasible to discriminate known ligands from randomly-chosen 'drug-like' molecules, for both antagonists and full agonists. Following this initial attempt, novel ligands have been identified for several GPCRs (dopamine D3 receptor, Neurotensin NTR-1 receptor, etc.).

In conclusion, virtual screening should be considered as a new weapon in our available arsenal. It is neither the *nec plus ultra* technology nor a method to only impress decision-makers. Depending on the environment, two major uses may be foreseen: (i) screening large commercially available compound collections for prioritizing experimental testing. This approach is mainly followed by either the academia or small companies that do not have the necessary infrastructure and financial capacity to screen corporate databases. (ii) screening commercial databases electronically, and 'in-house' libraries experimentally. This parallel approach is mainly of interest for big pharmaceutical companies that view VS as a complement to HTS.

However, as any computer-based technique, successful use of VS will entirely depend on the way it is utilized and the selection of sound experimental data to guide computing. It is, therefore, very important to teach this technique to students, stressing its incredible power but also its limitations.

References

[1] Mander, T. *Drug Discov Today* **5**, 223–225 (2000).

[2] Lahana, R. *Drug Discov Today* **4**, 447–448. (1999).

[3] Walters, W.P., Stahl, M.T., and Murcko, M.A. *Drug Discov. Today* **3**, 160–178 (1998).

[4] Pickett, S.D., McLay, I.M., and Clark, D.E. *J. Chem. Inf. Comput. Sci.* **2**, 263–272 (2000).

[5] Halperin, I., Ma, B., Wofson, H. and Nussinov, R. *Proteins* **47**, 409–443 (2002)

[6] Oprea, T.I. *J. Comput. Aided Mol. Design* **14**, 251–264 (2000).

[7] Lipinski, C.A, Lombardo, F., Dominy, B.W., and Feeney, P.J. *Adv. Drug. Deliv. Rev.* **23**, 3–25 (1997).

[8] Huuskonen, J., Rantanen, J., and Livingstone, D. *Eur. J. Med. Chem.* **35**, 1081–1088 (2000).

[9] Ertl, P., and Selzer, P. *J. Med. Chem.* **43**, 3714–3717 (2000).

[10] Zuegge, J., Schneider, G., Coassolo, P., and Lavé, T. *Clin. Pharmacokinet.* **40**, 553–563 (2001).

[11] Sadowski, J., and Kubinyi, H. *J. Med. Chem.* **41**, 3325–3329 (1998).

[12] Gillet, V.J., Willett, P., and Bradshaw J. *J. Chem. Inf. Comput. Sci.* **38**, 165–179. (1998).

[13] Wagener, M., and van Geerenstein, V.J. *J. Chem. Inf. Comput. Sci.* **40**, 280–292. (2000).

[14] Muegge, I., Heald, S.L., and Brittelli, D. *J. Med. Chem.* **44**, 1841–1846 (2001).

[15] Roche, O., Schneider, P., Zuegge, J., Guba, W., Kansy, M., Alanine, A., Bleicher, K., Danel, F., Gutknecht, E., Rogers-Evans, M., Neidhart, W., Stalder, H., Dillon, M., Sjogren, E., Fotouhi, N., Gollespie, P., Goodnow, R., Harris, W., Jones, P., Taniguchi, M., Tsujii, S., Saal, W., Zimmermann, G., and Schneider, G. *J. Med. Chem.* **45**, 137–142 (2002).

[16] McGovern, S., Caselli, E., Grigorieff, N., and Shoichet, B.K. *J. Med. Chem.* **45**, 1712–1722 (2002).

[17] Rishton, G.M. *Drug Discov. Today* **2**, 382–384 (1997).

[18] Leach, A.R. *J. Mol. Biol.* **235**, 345–356 (1994).

[19] Leach, A.R., and Lemon, A.P. *Proteins* **33**, 227–239 (1998).

[20] Sandak, B., Wolfson, H.J., and Nussinov, R. *Proteins* **32**, 159–174 (1998).

[21] Schnecke, V., and Kuhn, L.A. *Perspec. Drug. Discov. Des.* **20**, 171–190 (2000).

[22] Knegtel, R., Kuntz, I.D., and Oshiro, C.M. *J. Mol. Biol.* **266**, 424–440 (1997).

[23] Ewing, T.J.A., Makino, S., Skillman, A.G., and Kuntz, I.D. *J. Comput. Aided Mol. Design* **15**, 411–428 (2001).

[24] Oesterberg, F., Morris, G.M., Sanner, M.F., Olson, A.J., and Goodsell, D.S. *Proteins* **46**, 34–40 (2002).

[25] Claussen, H., Buning, C. and Rarey, M. *J. Mol. Biol.* **308**, 377–395 (2001).

[26] Rarey, M., Kramer, B., Lengauer, T., and Klebe, G. *J. Mol. Biol.* **261**, 470–489 (1996).

[27] International Human Genome Sequencing Consortium. *Nature* **409**, 860–921 (2001).

[28] Schafferhans, A., and Klebe, G. *J. Mol. Biol.* **307**, 407–427 (2001).

[29] Wojciechowski, M., and Skolnick, J. *J. Comput. Chem.* **23**, 189–197 (2002).

[30] e.g. the FRED docking programme (http://www.eyesopen.com/fred.html)

[31] Lorber, D.M., and Shoichet, B.K. *Prot. Sci.* 7, 938–950 (1998).

[32] Jones, G., Wilett. P., Glen, R.C., Leach, A.R., and Taylor, R. *J. Mol. Biol.* 267, 727–748 (1997).

[33] Morris, G.M., Goodsell, D.S., Halliday, R.S., Huey, R., Hart, W.E., Belew. R.K., and Olson, A.J. *J. Comput. Chem.* 19, 1639–1662 (1998).

[34] Taylor, R.D., Jewsbury, P.J., and Essex, J.W. *J. Comput. Aided Mol. Design*, in press

[35] Knegtel, R.M.A., and Wagener, M. *Proteins* 37, 334–345 (1999).

[36] Rarey, M., and Lengauer, M. *Persp. Drug. Discov. Des.* 20, 63–81 (2000).

[37] Hindle, S., Rarey, M., Buning, C., and Lengauer, T. *J. Comput. Aided Mol. Design* 16, 129–149 (2002)

[38] Mizutani, M.Y., Tomioka, N., and Itai, A. *J. Mol. Biol.* 243, 310–326 (1994)

[39] Welch, W., Ruppert, J., and Jain, A.N. *Chem. & Biol.* 3, 449–462 (1996).

[40] Lamarck, J.B. *Philosophie zoologique*, Flammarion Ed., Paris (1994).

[41] Clark, K.P., and Ajay, J. *J. Comput. Chem.* 16, 1210–1226 (1995).

[42] Taylor, J.S., and Burnett R.M. *Proteins* 41, 173–191 (2000).

[43] David, L., Luo, R., and Gilson, M.K. *J. Comput. Aided Mol. Design* 15, 157–171 (2001).

[44] Baxter, C.A., Murray, C.W., Clark, D.E., Westhead, D.R., and Eldridge, M.D. *Proteins* 33, 367–382 (1998).

[45] Hou, T. J., Chen, L., and Xu, X. *Protein Eng.* 12, 639–647 (1999).

[46] Goodsell, D.S., and Olson, A.J. *Proteins* 8, 195–202 (1990).

[47] Liu, M., and Wang, J. *J. Comput. Aided. Mol. Design* 13, 435–451 (1999).

[48] Trosset, J.Y., and Scheraga, H.A. *J. Comput. Chem.* 20, 412–427 (1999).

[49] Abagyan, R., Totrov, M., and Kuznetsov, D.J. *J. Comput. Chem.* 15, 488–506 (1994).

[50] Hart, T.N., and Read, R.J. *Proteins* 13, 206–222 (1992).

[51] McMartin, C., and Bohacek, R.S. *J. Comput. Aided Mol. Design* 11, 333–344 (1997).

[52] http://www.accelrys.com/insight/affinity.html

[53] Luty, B.A., Wasserman, Z.R., Stouten, P.F.W., Hodge, C.N., Zacharias, M., and McCammon, J.A. *J. Comput. Chem.* 16, 454–464 (1995).

[54] Stouten, P.F.W., Frommel, C., Nakamura, H., and Sander, C. *Mol. Simulation* 10, 97–120 (1993).

[55] http://www.schrodinger.com/Products/glide.html

[56] Sternberg, M.J.E., Gabb, H.A., and Jackson, R.M. *Curr. Opin. Struct. Biol.* 8, 250–256 (1998).

[57] Sobolev, V., Wade, R.C., Vriend, G., and Edelman, M. *Proteins* 25, 120–129 (1996).

[58] Burkhard, P., Taylor, P., and Walkinshaw, M.D. *J. Mol. Biol.* 277, 449–466 (1998).

[59] Miller, D.J., and Merz, Jr., K.M. *Proteins* 43, 113–124 (2001).

[60] Stahura, F.L., Xue, L., Godden, J.W., and Bajorath, J. *J. Mol. Graph. Model.* 17, 1–9 (1999)

[61] Su, A.I., Lorber, D.M., Weston, G.S., Baase, W.A., Matthews, B.W., and Shoichet B.K. *Proteins* 42, 279–293 (2001).

[62] Lamb, M.L., Burdick, K.W., Toba, S., Young, M.M., Skillman, A.G., Zou, X., Arnold, J.R., and Kuntz, I. *Proteins* 42, 296–318 (2001).

[63] Chen, Y.Z., and Zhi, D.G. *Proteins* 43, 217–266 (2001).

[64] Paul, N., and Rognan, D., personal communication.

[65] Kollmann, P.A. *Chem. Rev.* 93, 2395–2417 (1993).

[66] Tame, J. *J. Comput. Aided Mol. Design* 13, 99–108 (1999).

[67] Böhm H.-J. *J. Comput. Aided Mol. Design* 6, 593–606 (1992).

[68] Böhm, H.-J. *J. Comput. Aided Mol. Design* 8, 243–256 (1994).

[69] Eldridge, M., Murray, C.W., Auton, T.A., Paolini, G.V., and Lee, R.P. *J. Comput-Aided Mol. Design* 11, 425–445 (1997).

[70] Rognan, D., Laumoeller, S.L., Holm, A., Buus, S., and Tschinke, V. *J. Med. Chem.* 42, 4650–4658 (1999).

[71] Wang, R., Liu, L., Lai, L., and Tang, Y. *J. Mol. Model.* 4, 379–384 (1998).

[72] Wang, R., Lai, L., and Wang, S. *J. Comput. Aided Mol. Design* 16, 11–26 (2002).

[73] Gehlhaar, D.K., Verkhivker, G.M., Rejto, P.A., Sherman, C.J., Fogel, D.B., and Freer, S.T. *Chem. & Biol.* 2, 317–324 (1995).

[74] Muegge, I., and Martin, Y.C. *J. Med. Chem.* 42, 791–804 (1999).

[75] Gohlke, H., Hendlich, M., and Klebe, G. *J. Mol. Biol.* **295**, 337–356 (2000).

[76] Mitchell, J.B.O., Laskowski, R.A., Alex, A., and Thornton, J.M. *J. Comput. Chem.* **20**, 1165–1176 (1999).

[77] Ishchenko, A.V., and Shakhnovich, E.I. *J. Med. Chem.* **45**, 2770–2780 (2002)

[78] Stahl, M., and Boehm, H.-J. *J. Mol. Graph. Model.* **16**, 121–132 (1998).

[79] Charifson, P.S., Corkery, J.J., Murcko, M.A., and Walters, W.P. *J. Med. Chem.* **42**, 5100–5109 (1999).

[80] Bissantz, C., Folkers, G., and Rognan, D. *J. Med. Chem.* **43**, 4759–4767 (2000).

[81] Stahl, M., and Rarey, M. *J. Med. Chem.* **44**, 1035–1042 (2001).

[82] Wang, R., and Wang, S. *J. Chem. Inf. Comput. Sci.* **41**, 1422–1426. (2001)

[83] Clark R.D, Strizhev, A., Leonard, J.M., Blake, J.F., and Matthew, J.B. *J. Mol. Graph. Model.* **20**, 281–295 (2002).

[84] Terp, G.E., Johansen, B.N., Christensen, I.T., and Jørgensen, F.S. *J. Med. Chem.* **44**, 2333–2343. (2001)

[85] Hoffmann, D., Kramer, B., Washio, T., Steinmetzer, T., Rarey, M., and Lengauer, T. *J. Med. Chem.* **42**, 4422–4433 (1999).

[86] Wang, J., Kollman, P.A., and Kuntz, I.D. *Proteins* **36**, 1–19 (1999).

[87] Vieth, M., and Cummins, D.J. *J. Med. Chem.* **43**, 3020–3032 (2000).

[88] Paul, N., and Rognan, D. *Proteins* **47**, 534–545 (2002).

[89] Martin, Y.C. *J. Comb. Chem.* **3**, 1–20 (2001).

[90] Dixon, S.J. *Proteins. Suppl.* **1**, 198–204 (1997).

[91] Pearlman, D.A., and Charifson, P.S. *J. Med. Chem.* **44**, 3417–3423 (2001).

[92] Knegtel, R.M.A., Bayada, D.M., Engh, R.A., von der Saal, W., van Geerenstein, V.J., and Grootenhuis, P.D.J. *J. Comput. Aided Mol. Design* **13**, 167–183 (1999).

[93] Ha, S., Andreani, R., Robbins, A., and Muegge I. *J. Comput. Aided Mol. Design* **14**, 435–448 (2000).

[94] Keseru, G. *J. Comput. Aided Mol. Design* **15**, 649–657 (2000).

[95] Sotriffer, C.A., Gohlke, H., and Klebe, G. *J. Med. Chem.* **45**, 1967–1970 (2001).

[96] Ring, C.S., Sun, E., McKerrow, J.H., Lee, G.K., Rosenthal, P.J., Kuntz, I.D., and Cohen, F.E. *Proc. Natl. Acad. Sci. USA* **90**, 3583–3587 (1993)

[97] Li, S., Gao, L., Satoh, T., Friedman, T.M., Edling, A.E., Koch, U., Choski, S., Han, X., Korngold, R., and Huang, Z. *Proc. Natl. Acad. Sci. USA.* **94**, 73–78 (1997).

[98] Somoza, J.R., Skillman, A.G., Munagala, N.R., Oshiro, C.M., Knegtel, R.M.A., Mpoke, S., Fletterick, R.J., Kuntz, I.D., and Wang, C.C. *Biochemistry* **37**, 5344–5348 (1998).

[99] Tondi D., Slomczynska, U., Costi, M.P., Watterson, D.M., Ghelli, S., and Shoichet, B.K. *Chem. & Biol.* **6**, 319–331 (1999).

[100] Burkhard, P., Hommel, U., Sanner, M., and Walkinshaw, M.D. *J. Mol. Biol.* **287**, 853–858 (1999).

[101] Debnath, A.K., Radigan, L., and Jiang, S. *J. Med. Chem.* **42**, 3203–3209 (1999).

[102] Perola, E., Xu, K., Kollmeyer, T.M., Kaufmann, S.H., Prendergast, F.C., and Pang, Y–P. *J. Med. Chem.* **43** (401–408) 2000.

[103] Hopkins, S.C., Vale, R.D., and Kuntz, I.D. *Biochemistry* **39**, 2805–2814 (2000).

[104] Filikov, A.V., Mohan, V., Vickers, T.A., Griffey, R.H., Cook, P.D., Abagyan, R.A., and James T.L. *J. Comput. Aided Mol. Design* **14**, 593–610 (2000).

[105] Sarmiento, M., Wu, L., Keng, Y-F., Song, L., Luo, Z., Huang, Z., Wu, G.-Z., Yuan, A.K., and Zhang Z.-Y. *J. Med. Chem.* **43**, 146–155 (2000).

[106] Böhm, H.-J., Boehringer, M., Bur, D., Gmuender, H., Huber, W., Klaus, W., Kostrewa, D., Kuehne, H., Luebbers, T., Meunier-Keller, N., and Mueller F. *J. Med. Chem.* **43**, 2664–2674 (2000).

[107] Shapira, M., Raaka, B.M., Samuels, H.H., and Abagyan, R. *Proc. Natl. Acad. Aci. USA.* **87**, 1008–1013 (2000).

[108] Aronov, A.M., Munagala, N.R., Ortiz de Montellano, P.R., Kuntz, I.D., and Wang, C.C. *Biochemistry* **39**, 4684–4691 (2000).

[109] Aronov, A.M., Munagala, M.R., Kuntz, I.D., and Chang, C.C. *Antimicrob. Agents Chemother.* **45**, 2571–2576 (2001).

[110] Enyedy, I.J., Ling, Y., Nacro, K., Tomita, Y., Wu, X., Cao, Y., Guo, R., Li, B., Zhu, X., Huang, Y., Long, Y.-Q., Roller, P.P., Yang, D., and Wang, S. *J. Med. Chem.* **44**, 4313–4324 (2001).

[111] Grüneberg, S., Wendt, B., and Klebe, G. *Angew. Chem. Int. Ed.* **40**, 389–393 (2001).

[112] Schapira, M., Raaka, B.M., Samuels, H.H., and Abagyan, R. *BMC Struct. Biol.* **1**, 1 (2001).

[113] Joubert, F., Neitz, A.W.H., and Louw, A.I. *Proteins* **45**, 136–143 (2001).

[114] Waszkowycz, B., Perkins, T.D.J., Sykes, R.A., and Li, J. *IBM Systems J.* **40**, 360–376 (2001).

[115] Bressi, J.C., Verlinde, C.L.M.J., Aronov, A.M., Le Shaw, M., Shin, S.S., Nguyen, L.N., Suresh, S., Buckner, F.S., Van Voorhis, W.C., Kuntz, I.D., Hol, W.G.J., and Gelb, M.H. *J. Med. Chem.* **44**, 2080–2093 (2001)

[116] Doman, T.N., McGovern, S.L., Witherbee, B.J., Kasten, T.P., Kurumbail, R., Stallings, W.C., Connolly, D.T., and Shoichet, B.K. *J. Med. Chem.* **45**, 2213–2221 (2002).

[117] Powers, R.A., Morandi, F., and Shoichet, B.K. *Structure* **10**, 1013–1023 (2002).

[118] http://www.mdli.com/products/acd.html

[119] http://www.ccdc.cam.ac.uk/prods/csd/csd.html

[120] http://dtp.nci.nih.gov/docs/3d_database/structural_information/structural_data.html

[121] http://www.maybridge.com/

[122] http://leadquest.tripos.com/

[123] http://www.mdli.com/products/acdsc.html

[124] http://www.specs.net

[125] http://www.mdli.com/products/cheshire.html

[126] http://www.eyesopen.com/filter.html

[127] A set of 102 2D filters has been set-up and incorporated in a computer script by one of the authors (D.R.) to filter out unsuitable compounds from a flat database.

[128] Bissantz, C., Bernard, P., Hibert, M., and Rognan, D. *Proteins*, in press.

6
Scope and Limits of Molecular Docking

The aim of this chapter is to demonstrate that different docking problems need different docking procedures. Docking scenarios typically fall into one of the following categories: First, if the active site of the protein is not known, the search for both the binding site and, subsequently, the binding mode of the ligand is called a "blind docking" [1]. Methods of blind docking are also important for investigating protein-protein interactions. A special subcase of blind docking is the docking to homology models of a target where the position of the active site is assumed to be similar to the one in a template protein. Docking to models of transmembrane proteins, such as G-protein coupled receptors, falls into this category.

Second, if the site of the binding is known from X-ray diffraction or from NMR studies, the docking into the known active site is called a "direct docking". In this case, the focus of interest is the binding mode of the ligand. Several questions need to be addressed before starting a direct docking. It may be possible that:

(i) The active site is overlapping with the binding site of a cofactor. Should the cofactor be kept in the active site or not? Is the cofactor necessary for the binding of the inhibitors, knowing that for substrate binding the cofactor is essential?

(ii) The active site contains discrete crystal molecules of water, mediating the binding between the ligand and the target. Are they to be considered or ignored in the model?

(iii) The active site contains catalytic metal ions. How are they to be taken into account?

Third, recent developments in virtual screening techniques deal with factors which have been so far omitted in most of the docking approaches, namely the tautomeric or ionized states of the compounds under scrutiny, as well as the effect of the pH, pK or temperature micro-environment of the active site on ligand binding. Furthermore, induced fit and conformational changes of proteins must be taken into account. Finally, docking strategies are being developed for the situation when it is known that the putative inhibitor is expected to bind to the protein covalently. Figure 6.1 summarizes the mentioned problems and docking categories.

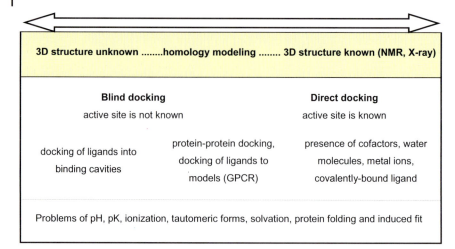

Fig 6.1   Categories of molecular docking.

6.1
Docking in the Polar Active Site That Contains Water Molecules – Viral Thymidine Kinase

In this section, we focus on the most prominent cases addressed in the above listing; direct docking and the role of water molecules in the active site. One of the characteristic features of an active site is the conformational change that occurs upon ligand binding and the subsequent rearrangement of water molecules. Discrete water molecules may mediate the binding of the ligand to the enzyme and may even directly participate in the catalysis. When running virtual screening assays, it is always a dilemma whether or not water molecules should be considered. For the screening of large databases the detailed binding mode of different compounds is usually unknown and hence this question is not easily decided. There are attempts, however, to take into account all the water molecules that contribute positively to binding [1–3]. If several water molecules are present, it may be hard to predict correctly how many of them participate in binding of the ligand and where their exact positions in the binding site are. Therefore, docking procedures usually recommend the removal of water molecules from the binding site. In most cases, the screening is thus performed for an empty active site devoid of water. However, in the heat-labile cholera toxin [3] and in HIV-1 protease [4], explicit water molecules play an important role in ligand binding.

The following example demonstrates the positive contribution of water molecules in the active site to an increased accuracy of docking, better prediction of the binding mode and consequently better results in screening. The target is the Herpes simplex virus 1 thymidine kinase (HSV1 TK). HSV1 TK is an important biomedical target and has been recently used in virus-directed enzyme prodrug gene therapy of cancer [5,6]. A specific characteristic of this protein is that it phosphorylates not only its natural substrate thymidine (dT) and certain pyrimidine analogs but also certain analogs of purine, for example the antiherpetic prodrug aciclovir (ACV) [7,8]. From crystal structure information it is known that the HSV1 TK active site is small and polar [9–14] and can be formally divided into two contiguous moieties. The first one accommodates the substrate and water molecules, the second moiety contains the cofactor ATP. Several ligands of HSV1 TK have been studied that are mainly analogs of pyrimidine and purine nucleic acid bases. Since experimentally measured affinities of many of them are known [11,15], these affinities can be compared with the calculated affinities predicted by the docking program.

6.1.1
Setting the Scene

The project described here is divided into three parts. First, in order to reproduce the crystallographically determined binding modes, known ligands are docked. The docking of the two main substrates thymidine and ACV is presented, and the effect of water molecules on the docking accuracy is studied. Next, 26 ligands of a small database of known binders to HSV1 TK are screened. Their predicted affinities were compared to the experimental values, always taking into account the possible presence of water molecules. This allows a detailed evaluation of the docking protocol exploitable for the ranking of the screened molecules. Finally, knowledge gained from the two preceding steps is transferred to the virtual screening of a large database yielding a new lead structure.

The coordinates of TK with dT and ACV were taken from the Protein Data Bank (2vtk and 2ki5) [12,15]. To determine the binding mode of thymidine and aciclovir the program AutoDock (version 3.0) was used. As described in detail in Chapter 5, AutoDock contains an efficient Lamarckian genetic algorithm for the docking of flexible ligands into a rigid binding site, which generates different low-energy docking poses (in modeling, a docking pose is often used as the synonym of a unique target-bound orientation and conformation of the compound). In this study, the poses have been sampled into clusters with rmsd values of less than 1 Å. AutoDock is not suitable for virtual screening, but it is a good tool for the analysis of individual binding modes. Considering its computer time requirements, 50 docking runs per compound can take around 1 hour on a SGI O_2 R5000 processor.

Docking of dT and ACV was performed by AutoDock into the thymidine-like active site which contains two essential water molecules, as well as into the ACV-like active site which is devoid of these water molecules. In AutoDock, as in many other docking programs, water molecules can be kept or positioned to the exact place in the binding site, thus being considered as a part of the protein. Parameters for water (position of hydrogens and partial charges of the atoms) can be defined by the user (e.g. taken from AMBER [16]). Thus, docking poses of both ligands were generated, while water molecules in the active site were kept or removed.

In the second part, docking of the database of 26 known ligands was performed using the program FlexX [17–19]. In contrast to AutoDock, this program contains an incremental construction algorithm, which first places a core fragment, and then peripheral fragments of the ligand into a rigid binding site (see Chapter 5). Based on this algorithm only one individual minimum energy pose is produced. FlexX uses formal charges and does not require the time-consuming *ab initio* charge calculation. FlexX is programmed mainly for screening purposes (Computer time requirements: around 3 min per compound on the above-mentioned workstation). In FlexX, water molecules can be defined in the active site in two ways; either selected from the FlexX menu "customize" as concrete water molecules or selected from the menu "the particle concept" [20]. Following this concept, water molecules are positioned within the site during the screening. In the study reported here, water molecules were kept at discrete, crystallographic positions.

In the third and final part of this project, a large database of 80 000 drug-like compounds from the Available Chemicals Directory (ACD) [21] was prepared and enriched by the 26 known enzyme ligands. The screening was performed first by using the program DOCK 4.0 [22, 23] and then by the FlexX program. DOCK matches ligands to the inverse image of the active site. Knowing about the importance of the two discrete molecules of water, two runs were performed: one with water present in the active site and the other one without water. Using DOCK water can also be treated as a part of the protein at a given position. DOCK is suitable for the screening of large databases and may serve as a pre-screening tool (see chapter 5). DOCK is a fast program, i. e. docking of one compound takes about 20 seconds on a SGI O$_2$ R5000 workstation. After the screening by DOCK, 1000 energetically top-ranked compounds from both runs were screened again using FlexX. For the ranking, the 26 known binders served as a reference. Top-ranked compounds were scored by five different scoring functions of the consensual scoring program CScore [24].

The clear advantage of the programs AutoDock, FlexX and DOCK, and the reason why they were employed in the study described here, is that they are widely used and lead to plausible results.

6.2
Learning from the Results

6.2.1
Water Contribution on dT and ACV Docking

Thymidine binds via a hydrogen bond network as shown in Figure 6.2. The presence of two water molecules mediating hydrogen bonds between the O(2) of the thymidine base and Arg176 is characteristic also for other pyrimidine derivatives [2,9–12,14]. ACV – a purine analog with an acyclic side chain mimicking the sugar moiety – binds in the active site in a manner similar to thymidine, yet via a different H-bond network. ACV does not interact via water-mediated hydrogen bonds. In analogy, the X-ray structures of other purine analogs reveal a binding pattern similar to that of ACV [11].

Before docking, the importance of the two discrete water molecules present in the thymidine-active site complex has been addressed. In approximately half of the crystal structures of viral thymidine kinases, these water molecules mediate the binding

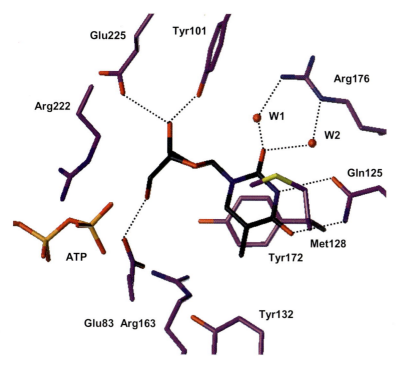

Fig. 6.2 Active site of HSV1 TK with bound thymidine. Thymidine is sandwiched between Tyr172 and Met128, creating five direct hydrogen bonds (dotted lines) and two hydrogen bonds mediated via two molecules of water, W1 and W2 (balls).

of the ligand to the protein. If pyrimidines are bound, those two water molecules are always present. It was decided, therefore, to run the docking twice: with the water molecules in the active site (water site) and without (empty site).

Docking of thymidine to the water site reproduced accurately the crystallographic binding mode, with rmsd values ranging from 0.36 to 0.72 Å for 50 runs. In contrast, docking to the empty site was less accurate, resulting in a fuzzy picture of the adopted docking poses (rmsd range increases to 0.74–1.23 Å). On the other hand, docking of ACV into the water site revealed several possible orientations and rmsd values compared to the X-ray orientation were not satisfying (rmsd > 1.2 Å). When the docking for ACV was run in absence of water, the prediction of the binding geometry improved and the docking procedure led to a single unique orientation (the rmsd value for the lowest-energy orientation decreased to 1.03 Å) (Figure 6.3). It can be concluded that, in contrast to thymidine which prefers docking to the water site, docking of ACV into the empty site leads to a more accurate prediction of the binding geometry. Similar results were obtained for other pyrimidine and purine analogs.

a

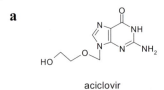

aciclovir

b

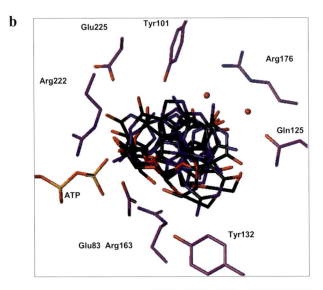

c

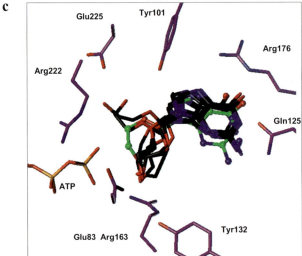

Fig. 6.3 Docking of aciclovir to HSV1 TK. (a) Chemical structure of aciclovir, (b) Docking clusters of aciclovir (C atoms in black color) into the HSV1 TK active site in the presence of water molecules (red balls) and (c) into the water-empty active site. Lowest-energy cluster of aciclovir in the water-empty active site is overlaying the ligand in its crystal structure binding mode (ball and sticks, C atoms in green color).

6.2.2
In Search of the Binding Constant

As already mentioned, a set of 26 ligands was docked using FlexX. In line with the preceding observations, the ligands were both docked reciprocally into the thymidine-like active site and into the ACV-like active site. Lowest-energy conformers have been selected and their affinities constants K_i have been calculated from Equation 1,

$$\Delta G_{FlexX} = RT \ln K_i, \tag{1}$$

where ΔG_{FlexX} stands for the free energy of binding calculated by FlexX. Table 6.1 shows in detail the results of K_i predictions for six ligands of the data set of 26 molecules. Comparing predicted K_i values to the K_i determined experimentally, it can be clearly seen that K_i values of pyrimidines improved by docking into the thymidine-like active site. A similar improvement can be observed for purine analogs by docking into the ACV-like active site. Hence, the presence of crystal water positively influences the prediction of binding constants of pyrimidines. In contrast, yet in full agreement with the known relevant crystal structures, an improvement of predicted values towards experimentally known affinity constants of purine analogs ACV and ganciclovir (GCV) is observed for the water-free situation.

Tab. 6.1 Prediction of affinity constants K_i of HSV1 TK ligands from the docking to water-containing or water-free active sites.

	Ligand	K_i exp. [μmol/l]	Docking in dT-like site (water is present)		Docking in ACV-like site (no presence of water)	
			ΔG [kJ/mol]	K_i predict [μmol/l]	ΔG [kJ/mol]	K_i predict [μmol/l]
Purine analogs	ACV	200	−19.8	341.0	−20.7	239.1
	GCV	47.6	−15.9	1631.3	−20.1	302.1
	DT	0.2	−35.3	0.7	−24.4	54.2
Pyrimidine analogs	BVDU	0.1	−33.1	1.6	−19.7	357.9
	IDU	0.09	−33.7	1.2	−23.9	64.7
	N-MCT	11.54	−26.4	23.5	−12.1	7586.7

GCV: ganciclovir, BVDU: bromovinyldeoxyuridine,
IDU: iododeoxyuridine, N-MCT: North-methanocarbathymidine

Differences of ΔG FlexX values for the 26 ligands, $diff\Delta G_{binding}$, were calculated by:

$$diff\Delta G_{binding} = \Delta G_{binding\ water\text{-}site} - \Delta G_{binding\ empty\ site} \tag{2}.$$

The results of this operation are displayed in the Figure 6.4. As expected, all compounds with negative values of $diff\Delta G_{binding}$ are pyrimidine analogs, whereas all

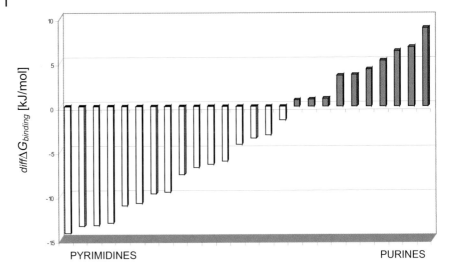

$$diff \; G_{binding} = \Delta G_{binding} \text{ for water-site docking} - \Delta G_{binding} \text{ for empty-site docking}$$

Fig. 6.4   Free Energy of binding profile for the docking into two different active sites of HSV1 TK.

compounds with positive values of $diff\Delta G_{binding}$ are purines. Docking purine deriva-tives into the water site obviously is disfavoured, and energy is gained if they are docked into the empty site. This is in full agreement with the predictions from dock-ing thymidine and aciclovir using AutoDock. It is encouraging to observe that the docking results unequivocally recognize the discrimination between the two main classes of analogs - pyrimidine and purine derivatives.

6.2.3
Application to Virtual Screening

Finally, the knowledge acquired about the impact of discrete water molecules on the binding of the ligands was transferred to the screening of 80 000 drug-like com-pounds from the ACD [21] database. The results led to diverse and interesting obser-vations when the 1000 top-ranked compounds were screened using the water site. One particular hit, which was ranked at position 41, was not found at all in the screening assay with the empty site. Experimental affinity assays, however, revealed that this compound indeed binds to HSV1 TK with a binding affinity in the sub-micromolar range, whereas thymidine and other pyrimidine inhibitors exhibit sig-nificantly lower affinities. This again shows the importance of taking water mole-cules into consideration for the screening and highlights the fact that discrete water molecules in the active site treated as a part of the docking target may lead to unique and powerful hits.

References

[1] Hetenyi, C. and Van Der Spoel, D. *Protein Sci.* **11**, 1729–1737 (2002).

[2] Pospisil, P., Scapozza, L., and Folkers, G. in: H.-D. Höltje and W. Sippl (Eds.) *Rational approaches to drug design: 13th European Symposium on Quantitative Structure-Activity Relationship*, Prous Science, Barcelona, 2001, pp. 92–96.

[3] Minke, W. E., Diller, D. J., Hol, W. G., and Verlinde, C. L. *J. Med. Chem.* **42**, 1778–1788 (1999).

[4] Lam, P. Y., Jadhav, P. K., Eyermann, C. J., Hodge, C. N., Ru, Y. et al. *Science* **263**, 380–384 (1994).

[5] Culver, K. W., Ram, Z., Wallbridge, S., Ishii, H., Oldfield, E. H. et al. *Science* **256**, 1550–1552 (1992).

[6] Bonini, C., Ferrari, G., Verzeletti, S., Servida, P., Zappone, E. et al. *Science* **276**, 1719–1724 (1997).

[7] Elion, G. B., Furman, P. A., Fyfe, J. A., de Miranda, P., Beauchamp, L. et al. *Proc. Natl. Acad. Sci. U. S. A.* **74**, 5716–5720 (1977).

[8] Keller, P. M., Fyfe, J. A., Beauchamp, L., Lubbers, C. M., Furman, P. A. et al. *Biochem. Pharmacol.* **30**, 3071–3077 (1981).

[9] Wild, K., Bohner, T., Folkers, G., and Schulz, G. E. *Protein Sci.* **6**, 2097–2106 (1997).

[10] Kussmann-Gerber, S., Kuonen, O., Folkers, G., Pilger, B. D., and Scapozza, L. *Eur. J. Biochem.* **255**, 472–481 (1998).

[11] Champness, J. N., Bennett, M. S., Wien, F., Visse, R., Summers, W. C. et al. *Proteins* **32**, 350–361 (1998).

[12] Bennett, M. S., Wien, F., Champness, J. N., Batuwangala, T., Rutherford, T. et al. *FEBS Lett.* **443**, 121–125 (1994).

[13] Perozzo, R., Jelesarov, I., Bosshard, H. R., Folkers, G., and Scapozza, L. *J. Biol. Chem.* **275**, 16139–16145 (2000).

[14] Prota, A., Vogt, J., Pilger, B., Perozzo, R., Wurth, C. et al. *Biochemistry* **39**, 9597–9603 (2000).

[15] Wild, K., Bohner, T., Aubry, A., Folkers, G., and Schulz, G. E. *FEBS Lett.* **368**, 289–292 (1995).

[16] Cornell, W. D., Cieplak, P., Bayly, C. I., Gould, I. R., Merz, K. et al. *J. Am. Chem. Soc.* **117**, 5179–5197 (1995).

[17] Rarey, M., Kramer, B., Lengauer, T., and Klebe, G. *J. Mol. Biol.* **261**, 470–489 (1996).

[18] http://cartan.gmd.de/flexx/

[19] Rarey, M., Kramer, B., and Lengauer, T. *Ismb* **3**, 300–308 (1995).

[20] Rarey, M., Kramer, B., and Lengauer, T. *Proteins* **34**, 17–28 (1999).

[21] http://www.library.wisc.edu/help/quick-guide/acd.htm

[22] Ewing, T. J. A., Makino, S., Skillman, A. G., and Kuntz, I. D. *J Comput. Aided Mol. Des.* **15**, 411–428 (2001).

[23] http://www.cmpharm.ucsf.edu/kuntz/

[24] http://www.tripos.com/software/cscore_print.html

7

Example for the Modeling of Protein–Ligand Complexes:
Antigen Presentation by MHC Class I

7.1
Biochemical and Pharmacological Description of the Problem

Cellular immunity is mediated by unique ternary complexes composed of major histocompatibility (MHC)-complex-encoded proteins, antigenic peptides and T lymphocytes. MHC molecules are glycoproteins. Their main function is to bind short antigenic peptides and present them to T lymphocytes at the surface of infected cells (Fig. 7.1).

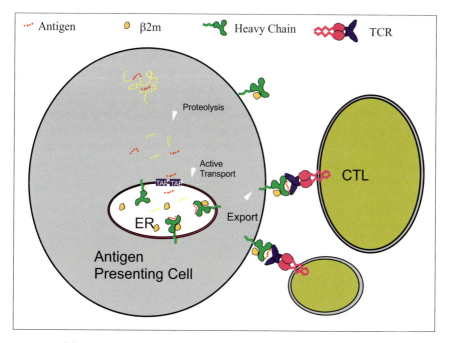

Fig. 7.1 Cellular immune response. CTL, cytolytic T lymphocyte; ER, endoplasmatic reticulum; TCR, T-cell receptor.

7.1.1
Antigenic Proteins are Presented as Nonapeptides

Hence, in contrast to the B lymphocytes, T lymphocytes do not recognize a protein antigen in its native conformation. The protein is normally processed inside the antigen-presenting cell and afterwards brought to the surface and bound to the MHC proteins. The MHC–peptide complex is then recognized by the T-cell receptor on CD8+ T lymphocytes. Most of the antigenic peptides for the class I MHC type are nonamers. This could be shown by elution of the peptides from purified MHC class I molecules. All peptides show conserved residues. Most of them have a conserved amino acid at position 2, which is believed to be the N-terminal anchor residue. Another conserved residue is the C-terminus which is hydrophobic in most cases and sometimes positively charged.

Amino acids at other positions are more variable and make either contact to the T-cell receptor in the ternary complex (T-cell receptor anchor residues) or should not play any decisive role in the formation of the MHC–T-cell interaction complex.

Until now, five MHC class I molecules have been crystallized. They are either bound to a mixture of peptides or to single peptides. Thus, the position of the ligands within the MHC molecules could be unambiguously determined and serves as a basis for the design.

It is a challenge for any design study that the presented antigenic peptides have been shown to be determinative for the whole process of the T-cell response. Length and sequence are the key features for starting the following biological responses:

- assembly and folding of the MHC proteins,
- binding to the MHC molecules,
- transport of this binary MHC–peptide complex to the cell surface,
- recognition of the binary complex by the T-cell receptor.

In terms of a subsequent modeling study, it is important to notice at this point, that:

1. Obviously no empty MHC molecules exist. Therefore, homology modeling of the protein alone does not make sense. This means docking of the ligand and model building of the binding site must take place in an iterative fashion.

2. Binding to the MHC molecule may be achieved by only two residues, namely at positions 2 and 9. This means that criteria have to be found for the discrimination of *good* and *bad* binders, as long as a ternary complex model taking into account the effects of the T-cell receptor, cannot be established yet.

7.1.2
Pharmacological Target: Autoimmune Reactions

Under normal circumstances, the immune system is self-tolerant. However, T-cell receptors which are normally selected to recognize only foreign peptide antigens bound to MHC molecules, may sometimes identify self peptides on MHC class I molecules. Obviously, only the ternary complex and not the MHC complex itself differentiates between self and non-self. T-cell receptors lacking the ability to differentiate between self and non-self may thus break the tolerance of the immune system and cause *autoimmune diseases.*

To date, special forms of arthritis are to our knowledge strongly linked to the expression of certain human leucocyte antigen molecules (MHC molecules). Presentation of bacterial proteins as antigenic peptides which remarkably resemble human self peptides, may be the molecular reason for the autoimmune disease.

In terms of medical treatment of the autoimmune diseases, blocking of the binding site of these special MHC molecules would at first glance be a highly attractive concept for a drug design study.

7.2
Molecular Modeling of the Antigenic Complex Between a Viral Peptide and a Class I MHC Glycoprotein

7.2.1
Modeling of the Ligand

The native ligands of MHC molecules are peptides. At the beginning of a drug design study one starts very often with the description of structural properties of the ligands. This activity is guided by the hope that structure–activity relationships might show up and facilitate the identification of the pharmacophore and/or the docking of the ligand into the binding site.

Peptides however, show considerable flexibility. They have a lot of local energy minima corresponding to a huge variety of different conformations. None of these may be associated with, or relevant for, the bound conformation at the MHC [1]. Furthermore, nothing is known to date about the structural features that determine the antigenic quality of the free peptides. And at last, as revealed by the biochemistry studies mentioned earlier, MHC protein folding seems to be a concerted action process with the binding of the peptidic ligand. The X-ray structures of the MHC complexes showed the bound peptides to have different binding geometries, ranging from an extended state to some sort of coiled geometry.

A set of synthetic peptides derived from the native nonapeptide Tyr-Pro-His-Phe-Met-Pro-Thr-Asn-Leu by subsequent truncation of the N- and C-terminus respectively provided a data basis for a preliminary structure–activity relationship study. A CoMFA study performed with eight peptides, truncated subsequently down from nona- to the pentapeptide, was based on the superimposition of helical geometries of the peptides. The study revealed the importance of the C-terminus to function as an anchor residue [2] (Table 7.1). The model explains the experimental findings by strong hydrophobic interactions of the C-terminus to a putative hydrophobic binding pocket at the MHC molecule. This information, however, might have been achieved by looking at the isolated C-terminal residues alone. Which relevance had then the helical conformation that had been used for superposition? None !

Tab. 7.1 Antigenic properties for cytolytic T lymphocytes (clone IE1)

Peptide name	Sequence	Recognition	Peptide concentration
Nona	YPHFMPTNL	+	10^{-9}
Nonar	PHFMPTNLG	+	10^{-3}
Nonal	MYPHFMPTN	−	
Octar	PHFMPTNL	+	10^{-7}
Octal	YPHFMPTN	−	
Heptar	PHFMPTN	+	10^{-7}
Heptal	YPHFMPT	−	
Penta	HFMPT	+	10^{-3}

The helical conformation had been taken because of its local stability. The idea arose from choosing appropriate starting conformations for a dynamic conformational analysis of the peptides. A second clue, that seduced us to accept helical conformations as the most plausible, came from the physical-chemists, who believed, that because of the helix dipole moment this conformation might be the most favored one for establishing protein–ligand interaction.

Both the theory about the importance of helix dipole moments for ligand interaction and the vacuum and solvent dynamics simulation of the isolated peptides, which showed the helical conformation to be the most stable, were found to be wrong in the light of the later occurring X-ray structures. However, for reasons of curiosity we had modeled in parallel protein–ligand complexes by energy-minimizing the different peptides bound to the MHC. It evolved that many more than only the helical conformations were preferred in the native environment.

Thus, the important lesson to learn was that peptides as substrates may be handled like other flexible molecules. The binding geometry is strongly case-dependent.

In the present case, some X-ray structures of MHC–ligand complexes, which had been published in the meantime, showed multiple nonapeptides bound to the active site. Their common structural features are two anchor residues. The binding geometry may additionally be markedly influenced by the third binding partner.

Thus, methods like the active analog approach [3], may fail in the case of evaluation of the docking geometries of peptidic ligands, although they have their profound merits in use with synthetic ligands.

This experience led to the decision, to find out as much as possible about the binding site. This knowledge, may it be experimental or theoretical, would help to restrain the degrees of freedom of the peptide's docking geometry.

The aforementioned advent of the first X-ray structures of the MHC class I molecules made it feasible to perform a homology modeling study. Sequences showed more than 70% homology, which should indicate a high degree of structural similarity in that class of proteins.

7.2.2
Homology Modeling of the MHC Protein

Affinity data of the peptides came from the H-2L^d receptor, a MHC-type protein but at present still unknown in structure. A X-ray structure of the human HLA-A2 MHC protein at 2.6 Å resolution was available [4] which shows 70% amino acid homology with the L^d molecule in the a$_1$ and a$_2$ domains (182 residues) of the peptide binding site.

7.2.2.1 Preparation of the Coordinates

In a first step, the crystal coordinates of HLA-A2 were refined in order to remove crystal packing effects. Three different types of calculation were performed with respect to the treatment of electrostatics:

1. A low dielectric model with distance-dependent dielectric functions.
2. A high dielectric model with dielectric constant set to 50 (D50 in Fig. 7.2).
3. A high dielectric model with explicit water molecules and a dielectric constant set to 1 (DW in Fig. 7.2).

The "best" structures (based on the deviation from X-ray backbone structure) were obtained by use of the high dielectric models. The models with the distance-dependent dielectric function very often overestimated salt bridges for instance between lysines and acidic amino acids, thereby creating non-regular structures. Thus, for a subsequent molecular dynamics simulation the distance dielectric model was dropped and only the "good" high dielectric models were used.

The molecular dynamics simulation procedures produced major discrepancies between the two starting structures of the model. The model with the dielectric constant set to 50 produced an unacceptably large deviation of almost 4 Å compared with 2 Å deviation obtained in the model with the explicit water molecules. Therefore, only the latter was found to provide a realistic picture at least near the solid-state geometry in the crystal and with minimized internal energy.

When inspecting the details of structure deformations in the vacuum dynamics simulation (model with dielectric constant set to 50) two prominent features could be seen to be responsible for the large rms deviation and typically occurring in vacuum simulation. Firstly, the active site, aligned by the two large helices shrank considerably by more than 50% (Fig. 7.2). Secondly, the helices themselves shrank by up to 6 Å. From this, a binding site resulted that would never be able to accomodate any ligand and thus was worthless for any further drug design procedure.

Part of the phenomenon can be explained by artificial hydrophobic collapses occurring with in vacuo simulations, because the hydrophobic surface attemps to become minimal. In contrast, by using of the explicit water model, the structure was seen to fluctuate around the X-ray-defined structure, giving the side chains the possibility of finding the optimal energy level with respect to a solvent environment. This led to a structural model averaged from 150 single structures during molecular dynamics simulation, that showed a backbone topology very near that of the X-ray backbone. Both the X-ray structure and its refined model by explicit treatment in a solvent are able to accomodate a nonapeptide as ligand in their binding site.

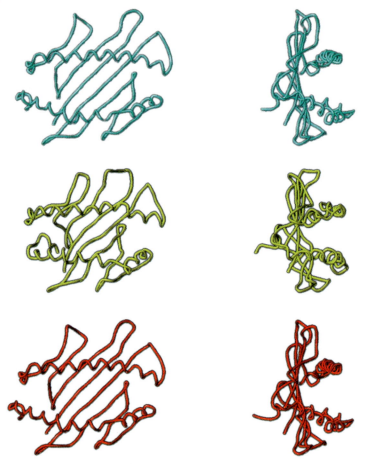

Fig. 7.2 Orthogonal views of three HLA backbone conformations: X-ray (top), D50 mean conformation (middle) and DW mean conformation (bottom).

7.2.2.2 Building the H-2L^d Molecule

The coordinates minimized in the explicit solvent environment were taken as a basis for the construction of the homology model of H-2L^d from HLA-A2. During the procedure, only the side chains were modified; the backbone was kept untouched. As has been described earlier, side chains were exchanged in a first step without taking care of interactions or optimal electrostatics. Because of differences in the sequence, a deletion occurred near the N-terminus.

This deletion was located in a loop structure. The latter is to be expected because at such a level of homology of both of the sequences, helices and sheets are always conserved. However, connecting loops are the positions, where individual substitutions occur in order to accomodate evolutionary fitting processes between different tissues or different species.

The loop identified between residues 12 and 18 had to be reconstructed from scratch. Because loops often have no ordered structure—or assume ordered structures only in the presence of a binding partner—we decided to perform a "loop search" in the Brookhaven crystallographic database in order to obtain an at least acceptable structure of the newly built loop. "Loop searches" perform a sequence alignment of the sequence to be built with the sequences already present in the protein data base (Fig. 7.3).

The algorithm is contained in most of the leading modeling packages. It presents the ten best "matches" from sequence comparison and subsequently cuts out the respective loop structures from the protein X-ray structures. The best fitting loop can be choosen to be built into the homology model, with respect to the distance of the N-terminal and C-terminal (see also section 4.3.3).

In the present case we found a loop with moderate homology but having a backbone geometry with only a 0.38 Å rms deviation from the N–C terminal distance defined by the template structure. At this stage the homology model represents a rough assembly of side chain orientations that must be refined in the subsequent steps.

The question is, whether this must be done in the presence of a docked ligand or with an "empty" binding site. According to the literature, as well as to our own experience, refinement of the complex should be performed preferentially with the ligand bound to the protein. However, the early steps might be done without any ligand, because the disorder of side chains may be too large to dock a ligand straight

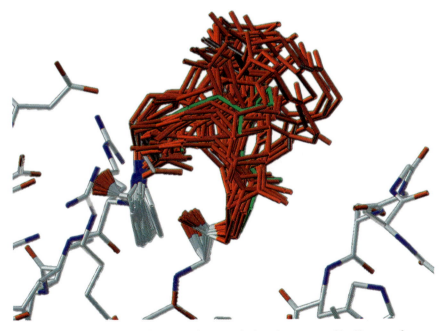

Fig. 7.3 Building a loop using the SYBYL_ loop search algorithm. Proposed backbone conformations are shown in red.

away in the binding site. The situation is worse in proteins that show an induced fit; in these cases multiple homology modeling steps are needed.

In the present case, we were sure that the MHC molecule was only folded correctly in the presence of the ligand. Therefore, we kept the backbone constant and removed steric interaction by energy-minimization. Subsequently the homology model was subjected to molecular dynamics simulation because we were curious to see whether it behaved like the X-ray structure; Indeed it did. Again, the model with the explicit water treatment showed a result much closer to the X-ray backbone, than to the model with dielectric constant set to 50. This analogy to the behavior of the X-ray structure gave us some confidence that the homology model obviously possesses at least some protein properties

As mentioned earlier, MHC molecules usually fold only in the presence of ligands; this led us to attempt a peptide docking in order to achieve the whole binary interaction complex and to proceed with the structure refinement of the complex. From previous QSAR studies it was suggested that the C-terminus of the ligand should bind to a hydrophobic environment or pocket. The peptides showed that at least positions 1 and 2 should additionally contribute to binding to the MHC molecule. We began to look for a binding pocket with hydrophobic properties that could accomodate the C-terminal amino acid of the peptide ligand and was limited in size, so as to exclude the amino acid tryptophan, which caused inactivity in the biological tests.

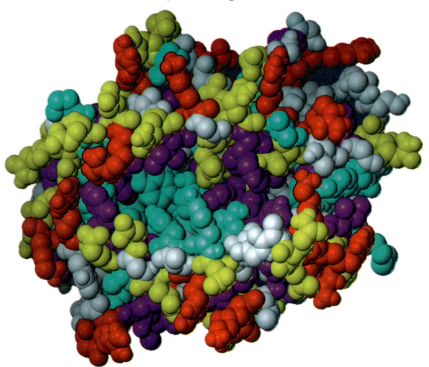

Fig. 7.4 Space-filling representation of the H-2L^d molecule with the Fauchère–Pliska hydrophobic scores. Color scheme: hydrophobic, magenta and cyan; hydrophilic, yellow and red.

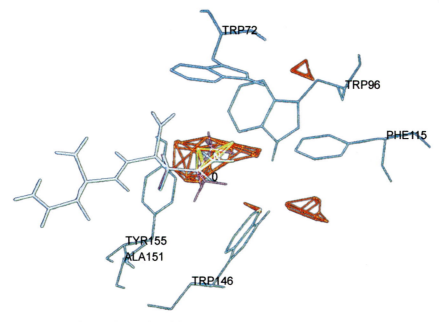

Fig. 7.5 Part of the binding pocket of the HLA C-terminus.

As a graphical aid, we used a hydrophobicity coloring scheme for the surface of our homology model of the H-2L^d molecule. The Fauchère–Pliska scale was applied to characterize hydrophobicity [5] (see Fig. 7.4). This particular scale was chosen because it has been determined experimentally and has already been successfully applied in studying antigenic sites in proteins. As expected, hydrophilic residues are located mainly at the surface of the protein, whereas hydrophobic areas are buried. One hydrophobic pocket, however, seemed quite large and suitably sized for the docking of the C-terminus. It consisted of three tryptophan, two phenylalanine and two tyrosine residues (Fig. 7.5).

Further indications came from experiments; two X-ray studies showed extra electron densities at the position of the hydrophobic pocket, supposedly resulting from co-crystallized peptides. The resolution, however, was not good enough to detail the interactions. This prompted us to choose the described pocket at the site for the C-terminal amino acid to serve as an anchor site for the peptide ligands. What of the remainder of the ligand's geometry? Residues at positions 1 and 2 had been predicted by QSAR and biochemically to be important also for the ligand's interaction with the MHC molecule, but nothing was known of the rest.

The helical interactions of the peptide ligand
Unfortunately we were seduced by the results of our conformational studies on the ligand. As mentioned earlier an α-helical structure was given to the ligands based on 3D–QSAR studies and for two additional reasons.

1. The α-helix turned out to be the most stable structure in solvent as predicted by molecular dynamics. However, this is the wrong line of enquiry. Minimization inside the binding site only provides useful information about the ligand, not energy minimization in vacuo or even in solvent. Nevertheless, we docked the ligand with helical geometry into the H-2L^d binding site. None of the currently known X-ray structures of MHC–ligand complexes show a regular helical conformation of the bound peptide. Therefore it is not likely that this was the only case of ligands being bound in such geometry. Strangely, manual docking provided an excellent interaction geometry for the nonapeptide Tyr-Pro-His-Phe-Met-Pro-Thr-Asn-Leu. The Leu9 fitted nicely into the hydrophobic pocket (Fig. 7.6(c)). The N-terminal Tyr1 interacted by aromatic interaction with Trp166 and by electrostatic interaction with Tyr56 of the MHC molecule (Fig. 7.6(a)). The second residue Pro2 could also be located nicely adjacent to Leu62 and Leu65 of the MHC, showing perfect

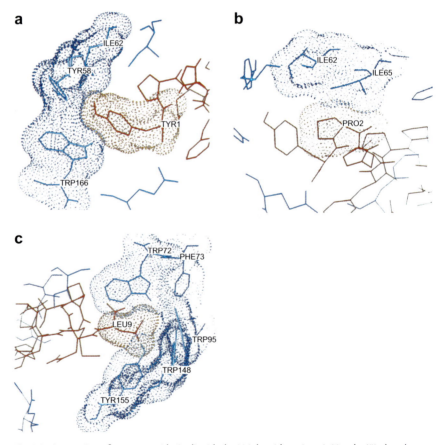

Fig. 7.6 Interaction of a nonapeptide (red) with the H-Ld residues (cyan). Van der Waals volumes of the nonapeptide's agretope residues (yellow) are represented embedded in the hydrophobic pockets: a) Tyr1; b) Pro2; and c) Leu9.

match of the solvent accessible area and hydrophobic interactions (Fig. 7.6(b)). Furthermore, the α-helix ideally spans the space between C- and N-terminal docking position.

2. Further support for this docking alternative was given by the fact, that in the present model, the nonapeptide interacted with highly polymorphic positions of MHC proteins, which could be interpreted as being a specific interaction. Of special note was the Ile62 and Ile65 contacting the proline at position 2, which is unique in these molecules.

This was a very optimistic view of the docking and ligand interactions. The opposite interpretation would be also possible, and is probably the more realistic. Although specificity is needed for ligand–MHC interaction, the main contacts are made to conserved residues in the class I MHC molecules. Therefore, the helical conformation that causes interactions of the ligand to non-conserved residues may be incorrect.

One may criticize that *any* model of antigen recognition must necessarily be incomplete as long as the contribution of the T-cell receptor (TCR) in the ternary MHC–peptide–TCR complex cannot be included in the simulation. Yet, to this argument, the fact must be recalled that the formation of the ternary complex is a stepwise process. First, the peptide must bind before the now-formed binary complex is subsequently recognized by the T-cell receptor. This again demonstrates the importance of integration of biochemical knowledge in the modeling process. Nevertheless we do not know, until the X-ray structure or model of the ternary complex with the T-cell receptor is available, which geometry is the real one. In conclusion, the resolution of the model was inadequate to provide a unique geometry for the protein–ligand and interaction complex.

The extended non-regular interaction of the peptide ligand

In a parallel study we examined the interaction of a peptide derived from influenza virus protein with the HLA-A2 MHC molecule that has been used as a template for the homology model of H-2L^d in the previous section [6].

Before docking the nonapeptide Gly-Ile-Leu-Gly-Phe-Val-Thr-Leu into the binding cleft of the HLA-A2 X-ray structure, the structure was truncated to the a_1 and a_2 domains. This approximation has been shown before not to alter significantly the 3D structure of a_1/a_2 because only limited interactions exists between a_1–a_2 domains and the a_3 and ‚-microglobulin domains. The latter are not suspected to contact antigenic peptides. The C-terminus was protected by a N-methyl group in order to avoid unrealistic electrostatic interactions. Furthermore, three crystalline water molecules have been placed in the binding site because they are visible in the X-ray structure and may be of importance for support of the peptide binding.

Subsequently, the nonapeptide Gly-Ile-Leu-Gly-Phe-Val-Phe-Thr-Leu was docked manually in the peptide binding groove. This time we used as a "template" an extra electron density map located in the binding groove of the MHC molecule that was observed in the X-ray structure, but could not be resolved to a unique ligand.

The crystallization of a mixture of peptides with the HLA-A2 molecule might have caused this extra electron density. Nevertheless, it provided a volume restraint into which the ligand should fit (Fig. 7.7). The N-terminal glycine was fixed at a

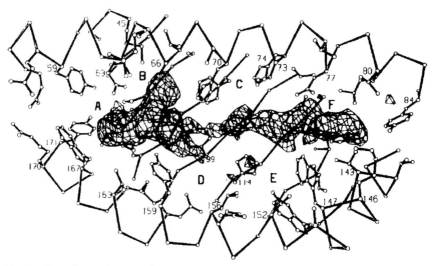

Fig. 7.7 Extra electron density in the X-ray structure of the HLA-A2 co-crystallized with a mixture of nonapeptides.

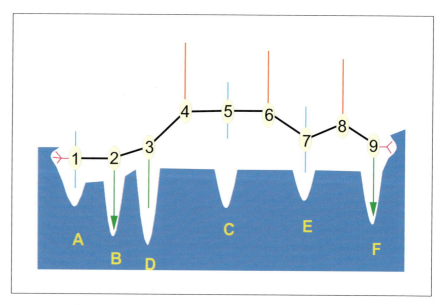

Fig. 7.8 Simplified model of nonapeptide binding to class I MHC proteins exemplified by HLA-B27. The six specificity pockets are labelled from A to F. N- and C-termini are colored in magenta. Main anchor side chains at P2 and P9 are displayed as arrows. MHC-binding side chains are colored in green, potential T-cell receptor (TCR)-binding side chains in red. Some amino acids (side chains in cyan) may bind both MHC and TCR molecules.

hydrogen bonding distance from the conserved residues Tyr7, Tyr59 and Tyr171 (Fig. 7.8). The second residue, isoleucine, is one of the formerly detected anchor residues, conserved among the peptidic ligands of MHC molecules. It has been placed such that its hydrophobic side chain contacted a set of three valines that form part of a hydrophobic pocket—which is also conserved among the MHC molecules. The peptide was further extended to the third and fourth residue by fitting it to the electron density map. Ile3 pointed in direction of a pocket named D, which was also hydrophobic in character aligned by two tyrosines and two leucines. Ile3 did not fill this pocket completely.

Again, this is a branching point in a modeling study that gives rise to two interpretations. Firstly, if there is a pocket it should be filled completely by the ligand's side chain in order to avoid "empty space" or large entropic contribution. It is unlikely that in the hydrophobic pocket water molecules fill the empty space. Secondly, the contrary argument is that evolution favors an optimal solution, not a maximal one. The ligand should be able to dissociate again. Filling every binding pocket to a maximum would considerably increase the energy necessary for dissociation and would decrease the variety in recognition of other peptides without losing specificity. Moreover, by an ultra high-affinity, a specificity would be gained that is probably not needed or that would even prevent the immune response.

Experimental constraints at branching points
At branching points the modeler needs help in deciding which branch to follow. Very often this information can be taken from previously known biochemical and pharmacological data. Therefore it is very important, as discussed earlier, to store and use all experimental information that can be accessed concerning the target protein and the ligands. In the present case the branching point provides an excellent opportunity to design a biochemical experiment that will prove the modeling process and the prediction, respectively.

Thus, we decided to design a non-natural peptide with maximum interaction at pocket D. From the synthetic ligand, we expected a much higher affinity to the MHC. (The design step will be described in detail in the next section.) In consequence we proceeded to model the natural ligands along the extra electron density template and let the pocket be only partly filled by Ile3. From there, the backbone turns upwards, directed to the opening of the binding cleft pointing to the solvent. For Gly at position 4 and Phe at position 5 a solvent or T-cell receptor interaction is hereby assumed. Val6, in contrast, is again contacting the MHC at a binding pocket, that is surprisingly polar, formed by two histidines, a threonine, and an arginine.

Positions 7 and 8 of the peptide ligand, following the electron density template, again point in the direction of the solvent. Interestingly, this is supported by a high variability at this position for the peptides eluted from the MHC complex in biochemical experiments.

Finally, position 9, bearing a leucine, was docked into the well-known hydrophobic C-terminal pocket, in line with details in the previous section regarding the H-2L^d molecule. This docking alternative created no steric conflicts and seemed rather reasonable from a chemical viewpoint. Although having a quite irregular, extended geometry, no violations of angles, torsion, etc. were identified on visual inspection of the interaction complex.

7.3
Molecular Dynamics Studies of MHC–Peptide Complexes

Modeling and docking of ligands has been described for H-2L^d and HLA-A2. In the next section the molecular dynamics simulations of two complexes will be discussed, namely HLA-A2 and—far more interesting from a viewpoint of unexpected results—HLA-B*2705 [6].

7.3.1
HLA-A2—The Fate of the Complex during Molecular Dynamics Simulations

For the HLA-A2 system, the homology modeling described in the previous section, the bimolecular complex, and the three crystalline water molecules were placed in a shell of water molecules. No periodic boundaries were applied, nor were positional constraints placed on the solvent atoms.

   As usual, the solvated complex was minimized first and subsequently subjected to a 100 ps molecular dynamics simulation at constant temperature. The system was coupled to a heating bath (see earlier). Analyses were taken from the last 60 ps. From the period of analysis (40–100 ps) a mean structure of the molecular dynamics simulation was obtained by averaging the atomic coordinates for 600 conformations. As shown in Fig. 7.9 the overall geometry is unchanged compared with the X-ray structure during molecular dynamics situation. The docking of the peptide did not significantly change the HLA-A2 structure. Most of the structural deviation—or even artefacts—were observed in loops connecting the α and , structures and the β-sheet, respectively. The more severe artefact was an unexpected flexibility of the a$_2$

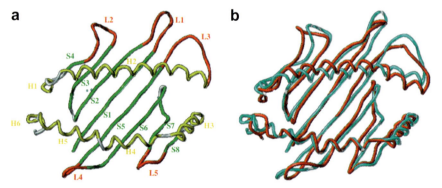

Fig. 7.9 Experimental and simulated 3D structures of HLA-A2.1. a) Crystal structure of the MHC protein. Only a1 and a2 domains are shown (α-helices H1–H6: yellow; ,-strands S1–S8: green; loops L1–L5: red; water: cyan balls). b) Superimposition of the crystal structure (cyan) and the time-averaged simulated conformation (red).

domain. This could be easily explained due to the lack of following a_3 and b_2 domains which are the native constituents of the complete MHC molecule, and are not present in the model. The general folding, however, was not disturbed because this movement was found oscillating around the X-ray structure.

A more detailed inspection of potential structure distortion was performed by using selected fits of secondary structures. This allows for distinguishing rigid body motions from distortion. In the first case, a structural element—for instance a helix—is translated or rotated as a whole. In a global fit, this would cause a bad rms value because of the average over the whole structure. If, however, only the structural element is fitted, it can be seen immediately that if the rms drops drastically, than the geometry is maintained and only the structure element has been translated or rotated. If, however, the rms value increases, a distortion of geometry is indicated.

Looking at the HLA-A2 step-wise by comparing secondary structure elements (Table 7.2) all but three secondary structure elements showed a considerably reduced rms, indicating that the overall geometry was maintained. The three elements that were more affected and showed an increased rms value are located near the C-terminus of the model. As explained earlier, the model takes into account only the binding site, consisting of a_1 and b_2 domains. The lacking a_3 and b_2 domains definitely do not contact any antigen, but might stabilize the whole MHC molecule. Therefore, if these domains are lacking just adjacent to a_2, some higher flexibility would be expected, probably resulting in distortion—exactly the situation that we found.

In an even more detailed step we traced the backbone angle variation. Around 80% of the ϕ and ψ angles did not alter for more than 20°. Only in the loop region were larger deviations found and, indeed, would have been expected. Interestingly, larger ϕ and ψ deviation which occurred at the C-terminal elements were always compensated by the next ϕ and ψ angles along the sequence, thus maintaining perfectly the overall secondary structure and the interaction geometry.

To analyze observations and statistics of atomic fluctuations is again one level deeper in a detailed study of the system. All fluctuations correlated with the motions of larger parts and substructures described earlier.

Atomic fluctuation has been analyzed in the present case, especially for water molecules, the whole system being surrounded by some 1300 water molecules with TIP3P potentials.

Consideration of atomic fluctuation of water oxygen atoms is one method of analyzing the quality of the molecular dynamics simulation. We could clearly detect four different types of water: first, the explicit water molecules inside the binding cleft showing participation in the hydrogen network of the ligand; second, water molecules bound to the surface of the interaction complex, exhibiting fluctuations only; third, the bulk water with moderate variation; and fourth, water molecules at the water–vacuum edge being the most flexible and showing rms values over 1.0.

Tab. 7.2 Root mean square (rms) deviation from the crystal structure. Coordinates of the HLA-A2 a1–a2 domains were time-averaged over 40–100 ps and compared with the crystal structure. Deviations in nm have been calculated for backbone atoms after fitting the whole structure (rms 1) or selected sequences (rms 2)

Structure	Position	rms1	rms2
a1–a2 Domains	1–182	0.182	0.182
a1 Domain	1–90	0.167	0.162
a2 Domain	91–182	0.195	0.191
secondary elements		0.166	0.078
α-Helices		0.173	0.091
H1	50–53	0.206	0.128
H2	57–84	0.149	0.099
H3	138–150	0.138	0.033
H4	152–161	0.091	0.043
H5	163–174	0.211	0.137
H6	176–179	0.371	0.042
β-Strands		0.120	0.066
S1	3–12	0.074	0.074
S2	21–28	0.073	0.043
S3	31–37	0.083	0.043
S4	46–47	0.122	0.024
S5	94–103	0.076	0.056
S6	109–118	0.076	0.122
S7	121–126	0.229	0.074
S8	133–135	0.185	0.044
Loops		0.250	0.106
L1	13–20	0.226	0.137
L2	38–45	0.264	0.128
L3	85–93	0.175	0.131
L4	104–108	0.348	0.024
L5	127–132	0.201	0.057
Crystal water	193–195	0.249	0.132

7.3.2
HLA–B*2705

For the HLA-B*2705 MHC molecule a X-ray crystal structure was available [7] from which the coordinates were taken. The peptides bound to the MHC molecules were derived from the nonamer which had been modeled into the binding cleft of the X-ray structure; this had the sequence Ala-Arg-Ala-Ala-Ala-Ala-Ala-Ala-Ala. The other peptides have been created simply by replacing the alanines subsequently by the corresponding residue of the desired derivative. Its side chains were centered in the binding-pockets according to the electron density map of the peptide. Those residues responsible for the interaction with the T-cell receptor pointed towards the solvent as no receptor interaction could be taken into account. The backbone geometry of the crystal structure was taken as a template for all nonapeptides bound to the MHC [8].

So far, the entire situation is quite similar to the complex described earlier, so, what is interesting about B*2705? The B*2705 binding motif has been characterized by analysis of a variety of the bound peptides. These were eluted from the native complex by HPLC and sequenced. From the set of peptides available, position 2 was identified as a main anchor residue, always being an arginine. The other anchor positions are 1, 3, and 9, preferring hydrophobic and positively charged residues, and 2 and 9 being the most important. These experimental data do not, however, entirely account for the HLA-B*2705 binding properties of several bacterial peptides.

Peptides from *Chlamydia trachomatis* could be shown to bind to HLA-B*2705. They stem from the 57 kDa heat shock protein of *C. trachomatis* and are probably responsible for an anti-immune reaction causing diseases related to rheumatoid arthritis. The bacterial peptide Leu-Arg-Asp-Ala-Tyr-Thr-Asp-Met-Leu, for example, fits nicely the consensus sequence. Arginine in position 2 and a hydrophobic or positively charged amino acid in positions 1, 3, and 9 represent a binding pattern as defined by the anchor residues, except for position 3. Thus, the peptide is expected to show affinity to HLA-B*2705, but does not because it is not recognized. The opposite is true for the peptide Arg-Arg-Lys-Ala-Met-Phe-Glu-Asp, i.e. an octapeptide rather than a nonapeptide, but with the only similarity being the arginine in position 2. If this position is correctly docked into the second pocket, the peptide would be too short for any interaction with the hydrophobic pocket at position 9, this being very important for stabilization of the nonapeptides. Surprisingly, the octapeptide is recognized by the MHC molecule, although its binding motif does not match the experimental pattern very well.

These were the reasons why we were interested in the B*2705 complex and tried to rationalize the structure–activity relationships by performing molecular dynamics simulations. Six different peptides were chosen for this study, the rationale for choice being the following (see Table 7.3):

1. The nonapeptide Arg-Arg-Ile-Lys-Ala-Ile-Thre-Leu-Lys has been described as part of the crystal structure. Therefore this peptide served as a basis for the setup of appropriate parameters for the molecular dynamics simulations. If the X-ray structure could be reproduced by a certain set of molecular dynamics parameters, we would be willing to accept these conditions for the whole series of peptide–HLA complexes to be simulated. We are fully aware, that this assumption is in fact an extrapolation. It is, however, the most cautious one that can be done in this case.

2. The second (Glu-Arg-Leu-Lys-Glu-Ala-Ala-Glu-Lys) and third (Arg-Arg-Lys-Ala-Met-Phe-Glu-Asp-Ile) peptides were taken as positive controls, because they showed high-affinity binding. The sequence Glu-Arg-Leu-Ala-Lys-Leu-Ser-Gly-Gly has been taken as a negative control, as it does not bind to the B*2705.

Both peptides mentioned previously (Arg-Arg-Lys-Ala-Met-Phe-Glu-Asp and Leu-Arg-Asp-Ala-Tyr-Thr-Asp-Met-Leu) were used as test cases, where we hoped to be able to explain the unexpected binding properties.

For the docking of the octapeptide we had to accept some compromises. The negatively charged Asp at the C-terminus was certainly not expected to interact with the pocket for the normal C-terminal residues of the nonapeptides because the pocket itself is aligned by two Asp residues. Thus, the docking was based on the hypothesis that the Asp might be able to simulate the normal C-terminus of the nonapeptide. Therefore, the octapeptide was docked without having a side chain interaction of pocket F, which is normally responsible for binding the C-terminus of the nonapeptide. The more extended conformation of the octapeptide could be accomplished by moderating the bulge, normally occurring between position 4 and 7 and in reality supposedly binding to the T-cell receptor (Fig. 7.10).

Tab. 7.3 Binding of bacterial peptides to HLA-B*2705

Peptide no.	Sequence	Origin	Binding
1	Arg-Arg-Ile-Lys-Ala-Ile-Thr-Leu-Lys	Model	not determined
2	Gln-Arg-Leu-Lys-Glu-Ala-Ala-Glu-Lys	Hsp 75[a]	good
3	Arg-Arg-Lys-Ala-Met-Phe-Glu-Asp-Ile	Hsp 57[b]	best
4	Glu-Arg-Leu-Ala-Lys-Leu-Ser-Gly-Gly	Hsp 57	non
5	Leu-Arg-Asp-Ala-Tyr-Thr-Asp-Met-Leu	Hsp 57	non
6	Arg-Arg-Lys-Ala-Met-Phe-Glu-Asp	Hsp 57	good

[a] From *Escherichia coli.*
[b] From *Chlamydia trachomatis.*

7.3.2.1 The Fate of the Complex during Molecular Dynamics Simulations

Here we describe only the main concepts used to distinguish between "good" and "bad" binders. The detailed analysis with listings of every hydrogen bond interaction may be duplicated in the original papers [8, 9].

In fact, the molecular dynamics simulation proved able to account for anomalous binding of the bacterial peptides. Again, as shown previously, the most important criteria for the judgment of the models were hydrogen bonding, solvent-accessible areas, and atomic fluctuations. To begin with the latter, we were mainly interested in the behavior of the binding pockets related to anchor residues 2 and 9. When bound to the inactive peptides Glu-Arg-Leu-Ala-Lys-Leu-Ser-Gly-Gly and Leu-Arg-Asp-Ala-Tyr-Thr-Asp-Met-Leu, respectively, atomic fluctuations were increased dramatically compared with the native peptides. As expected, the atomic motions of the pockets

▶

Fig. 7.10 Time-averaged conformation of HLA-B27 in complex with six peptides (A–F). The backbone atoms of the two α-helices delimiting the peptide-binding groove are displayed here with the side chains of peptide-binding residues. The C^2 positions of bound peptides (in bold) are labeled from P1 (N-terminus) to P9 (C-terminus). Only the peptide anchor side chains are shown. MHC–peptide hydrogen bonds are represented by broken lines and water molecules by balls.

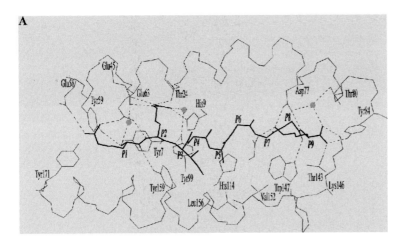

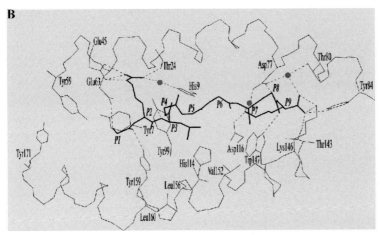

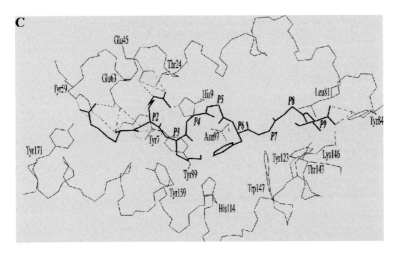

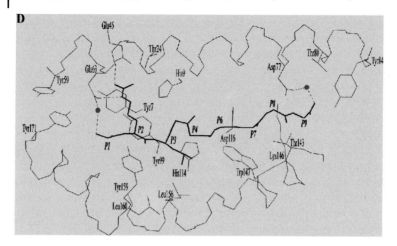

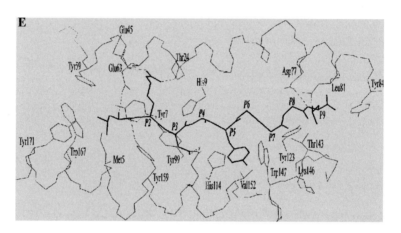

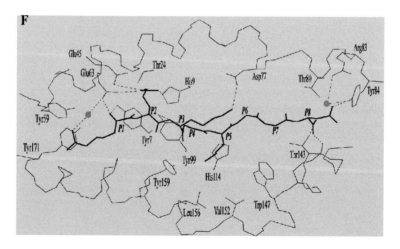

correlate clearly with the nature of side chain of the peptides. Good binders have perfectly complementary side chains properties. Thus, inactive peptides lack side chain interactions or show only weak interaction with pockets 2 and 9. This results in an increased atomic mobility. Logically, a similar pattern emerged for the analysis of hydrogen bonding in the peptide–MHC interaction. Again, our main interest focused on the binding pockets for residues 2 and 9.

For the native complex (X-ray structure) from 15 hydrogen bonds interactions at positions 2 and 9, all but four could be reproduced by the molecular dynamics simulation. This is quite a lot if one considers the highly reductionistic model. Interestingly, these four missing H bond interactions compared with crystal structure could be shown to be engaged in the water intercalation effect at the N-terminus. Water molecules slowly "walk in", starting at the N-terminus, slightly loosening the side chains from their binding pockets. Although this might be an artefact of the simulation, the principle reflects the differences between solution and crystal state.

A dramatic drop of H bond interactions is seen for the inactive peptides; this was also expected from atomic fluctuation analysis with only two of seven H-bonds remaining. For both peptides, the C-termini have lost their original H-bonds, while at the N-terminus the peptides are not hydrogen bonded at all.

The most interesting situation is that for the octapeptide. Thirteen H bonds emerge after and during 150 ps molecular dynamics simulation. The main anchor residue arginine resides in pocket B, its native location. The middle part of the peptide (residues 4–7) does not interact at all with the MHC molecule. The C-terminus, however, is in fact replacing the normal carboxyl end of the nonapeptide (Fig. 7.10).

7.4
Analysis of Models that Emerged from Molecular Dynamics Simulations

Four criteria have been used to analyze the binding situation of ligand–protein complexes and to correlate them at least non-quantitatively to the experimental observations. The criteria were: hydrogen bonding networks, interaction energies, solvent-accessible surface, and atomic fluctuations.

Attempts to quantify the results of the molecular dynamics simulation were very difficult. Therefore, the use of calculated interaction energies may be the weakest part of the four criteria mentioned. Simplification in quantification of electrostatic interaction and hydrophobic binding, respectively, will provide only rule-of-thumb values for estimation of ligand–protein interactions. Only in rare cases does the reductionistic nature of the models allow for a *quantitive* structure–activity relationship based on interaction energies. Thermodynamic analyses of ligand–protein interaction are currently under study and may be used in future to calibrate calculated interaction energies. Furthermore, refined approaches to calculate electrostatics—as designed by the use of the Boltzmann–Poisson equation [10]—may be helpful in the detailed quantitative analysis of interaction energies.

7.4.1
Hydrogen Bonding Network

In the studies presented, hydrogen bond properties have been described in terms of a donor (D)–acceptor (A) distance lower than 0.35 nm and a D–H–A bond angle value of 120–180°. Time-averaged conformations of up to 200 ps simulation time were analyzed in most cases.

In general, the total number of MHC–peptide hydrogen bonds was strictly correlated to the binding properties of the corresponding peptide. In all simulations the significant pattern of hydrogen bonding networks could be reproduced for crystal structures. For the non-binders—peptides exhibiting low experimental affinities—a dramatic loss of hydrogen bonds could generally be observed. This was especially true for the N- and C-termini. The effect was less dramatic for the main anchor position 2.

Differences in the hydrogen bond pattern, when compared with the crystal structures, may also occur with high-affinity ligands, for example, water molecules moving in at C- and N-terminal binding pockets, though whether this is an artefact or simply shows an early step of dissociation remains unclear.

In our opinion molecular dynamics simulations represent much of a solvated state. Furthermore, the molecules are provided with kinetic energy which enables them to find new positions, not necessarily tracing down to the global minimum. Thus, differences to the crystal state might be expected which—if they occur coincidentally at the termini of the bound peptides—may cause the molecular dynamics simulations to reflect something of the reality of dissociation behavior of the ligand–protein complexes. In the present case of ligand–MHC interaction, careful analysis of the hydrogen bonding pattern was the basis for predicting correctly these parts of the ligands that could be replaced by non-interacting spacer residues (see the next section).

Table 7.4 shows how such a H-bond pattern emerging from the molecular dynamics simulation can be represented. Low-affinity binders (peptides 4 and 5) can be detected directly by loss of H-bond interactions compared with the crystal structure of a native ligand bound to the MHC (first column).

7.4.2
Atomic Fluctuations

The atomic fluctuations were computed and compared with the fluctuations from the crystallographically determined temperature factors. This allows for an illustration of the relative gain or loss in flexibility compared with the native X-ray structures.

Atomic fluctuations were found to be an excellent tool which provides direct insight in the activity-correlated properties of the ligands, as they depend directly on strong or weak electrostatic and/or hydrophobic interactions. Graphical representation facilitates the direct comparison of several ligands with respect to their binding properties within the same scale. This is illustrated by the following example (Fig. 7.11).

The upper graph in Fig. 7.11 represents the atomic fluctuations of the binding pockets. Fluctuations are calculated from time-averaged conformations of the last 500 conformations of the molecular dynamics simulation. The binding pockets

Tab. 7.4 MHC-peptide hydrogen bonds. Peptide positions (Pn) are labelled from 1 (N-terminus) to 9 (C-terminus). Closed and open boxes indicate the presence or absence of a peptide–MHC hydrogen bond, respectively (time-averaged distance between donor D and acceptor A less than 3.2 Å, D–H··· and 180°). Crosses indicate the absence of specific side chains for some peptides. Hydrogen bonds have been analyzed for the crystal structure (X-ray) and during the last 50 ps of the simulation over 500 HLA-peptide conformations for each complex with peptides 1 to 6 (MD1 to MD6)

Peptide	HLA*B2705	X-ray	MD1	MD2	MD3	MD4	MD5	MD6
P1(N)	Tyr7(OH)	■	■	□	■	□	□	■
	Tyr59(OH)	□	□	■	■	□	□	■
	Glu63(OE1)	□	□	■	□	□	□	■
	Glu63(OE2)	□	□	□	□	□	■	■
	Tyr171(OH)	■	□	□	□	□	□	□
P1(NE)	Glu163(OE2)	■	□	×	□	×	×	□
P1(NH1)	Glu63(OE2)	□	■	×	■	×	×	□
P1(NH2)	Glu58(OE2)	□	□	×	□	×	×	□
	Glu166(OE2)	□	□	×	□	×	×	□
P1(O)	Tyr99(OH)	□	□	□	■	□	□	■
	Tyr159(OH)	■	■	□	□	□	■	□
P2(N)	Glu63(OE1)	■	■	■	■	□	□	■
P2(NE)	Glu45(OE1)	□	■	□	■	□	□	□
	Glu45(OE2)	□	□	■	□	□	□	□
	Glu63(OE2)	□	□	□	□	■	■	□
P2(NH1)	His9(NE2)	■	■	□	■	□	□	■
	Thr24(OG1)	■	■	□	□	□	■	■
	Glu45(OE1)	□	■	□	□	□	□	□
P2(NH2)	His9(NE2)	■	■	■	■	□	□	■
	Glu45(OE1)	■	■	■	□	□	□	□
	Glu45(OE2)	■	■	■	■	□	□	□
P2(O)	Arg62(NH1)	■	■	□	□	□	□	□
P3(N)	Tyr99(OH)	□	□	□	■	□	■	□
P3(OD2)	Gln155(NE2)	×	×	×	×	×	■	×
P3(NZ)	Asp77(OD1)	×	×	×	×	×	×	■
P8(O)	Lys146(NZ)	■	■	■	□	□	□	■
	Trp147(NE1)	■	■	■	■	□	□	□
P8(OXT))	Thr143(OG1)	×	×	×	×	×	×	■
P8(OD2)	Tyr84(OH)	×	×	×	×	×	×	■
P9(N)	Asp77(OD1)	■	■	□	□	□	■	×
P9(NZ)	Asp77(OD2)	□	■	■	×	×	×	×
	Asp116(OD2)	■	■	□	×	×	×	×
P9(O)	Tyr84(OH)	□	□	□	■	□	□	×
	Thr143(OG1)	■	□	□	□	□	□	×
	Lys146(NZ)	■	□	□	□	□	□	×
P9(OXT)	Tyr84(OH)	■	□	■	□	□	□	×
	Thr143(OG1)	■	□	□	■	□	□	×
	Lys146(NZ)	□	□	■	■	□	□	×
Total		15	18	16	13	2	7	13
Backbone		10	9	9	8	1	4	6
Side chains		5	9	7	5	1	3	7

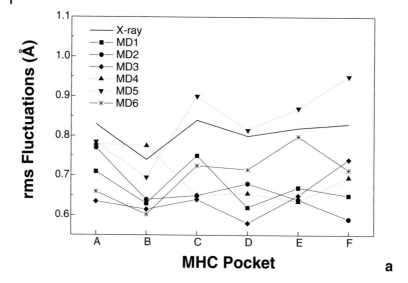

a

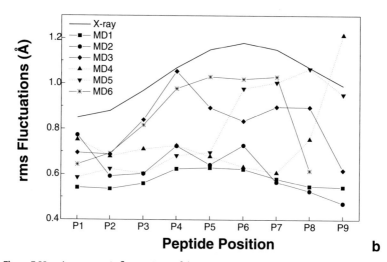

b

Figure 7.11 a) rms atomic fluctuations of the six HLA-B27 specificity pockets in complex with the bacterial peptides. The values for the X-ray structure (bold line) were obtained from temperature factors. b) rms atomic fluctuations of HLA-B27-bound peptides (backbone atoms) for the crystal structures and six MD conformations.

themselves are formed by up to six side chains. If one looks at the active peptides, which means high-affinity binders, pockets B for residue 2, pocket D for residue 3, and pocket F for residue 9, show the lowest fluctuations. The second amino acid of every peptide is bound to pocket B with such a high affinity, that the movement of the side chains of the pockets is dramatically restrained. This is represented by the solid lines clustering around 0.65 Å rms fluctuations.

Switching to the low-affinity peptides (the non-binders), the situation changes completely. The residues forming pocket B show rms fluctuations between 0.7 and 0.8 Å. This indicates a larger flexibility of the binding pocket and, vice versa, a less tight binding or only few interactions from the peptide to the binding sites. The situation is much more dramatic for specificity pocket F, which normally binds residue 9. There is a difference of nearly 0.4 Å in atomic fluctuations, which indicates a large movement of the pocket and hence no interactions to the peptide's C-terminal residue. In summary, it can be seen that the highest fluctuation values are found for the complex with inactive peptides.

The lower graph in Fig. 7.11 provides a complementary picture, showing the fluctuation of the binding ligands, the peptides. The graph is even easier to interpret; low atomic fluctuations indicate tight binding, and vice versa. Again, the most active peptides 1, 2, and 3, show the lowest fluctuations.

Much more interesting in this context is the behavior of the octapeptide. This is also an active peptide, and thus needs tight binding to the MHC molecule in order to be presented to the T-cell receptor. The octapeptide (*) shows a highly fluctuating sequence in the middle, namely for the residues 4 to 7. Amino acids at position 1, 2, and 8, however, are at very low fluctuation levels.

Thus, the octapeptide reveals the importance of the binding pockets for peptide presentation, the binding to pockets A, B, and F being complementary to residues 1 and 2, while the C-terminus of the ligands seems the precondition for presentation to T-cell receptor.

7.4.3
Solvent-Accessible Surface Areas

For peptide side chains binding to a pocket within the surface of a receptor, the idea of correlating this process with the residual surface that is accessible to the solvent, seems straightforward. As we learned from the X-ray studies [7], binding pockets B and F that bind residues 2 and the C-terminus 1 of the peptidic ligands really bury these side chains. Thus, most of the side chain-solvating molecules must be removed and replaced by an interaction to the side chains of the receptor protein that make up the walls of the binding pocket. The residual surface that is still accessible to the solvent after the ligand has been docked, is a measure of the depth of binding into the pocket. It correlates with the tightness and more or less with the strengths and number of binding interactions with the pocket.

Accessible and buried surface areas, respectively, were computed using an algorithm from Connolly [11]. A probe atom with a 1.4 Å radius is used to walk around the ligand or the part that is visible. The radius of the probes simulates a water molecule. In order to quantify the results in terms of percentages, free peptides of the sequence Gly-Xaa-Gly were built in an extended conformation and computations performed with those in a similar fashion. These served as an example of fully solvated or fully accessible reference surfaces.

Using this description technique, the important binding pockets of the MHC molecules could be easily identified by analyzing the ligand–protein complexes.

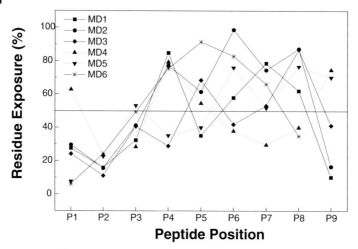

Fig. 7.12 Solvent-accessible surface area of MHC-bound nonapeptides. All values were computed for relaxed molecular dynamics mean structures, time-averaged over 500 conformations during the last 50 ps of the molecular dynamics simulation.

Fig. 7.12 shows the results of the study on B*2705–ligand interaction. Again, the active peptides are represented by solid lines, the inactive ones by open lines; the octapeptide is represented by *. The horizontal line in the graph represents a 50%-buried residue. It is immediately clear from the graph that residues at position 1, 2, and 9 are the important binding locations of the MHC molecule, because they are buried almost completely. Significantly this is not true for the low-affinity peptides, since their side chains, even at the C-terminus, are 70% exposed to the solvent. Again, the graph provides an immediate interpretation of the pharmacophore of the peptidic ligands. Residues that bind to MHC can be identified by their high degree of burying; inactive peptides can be seen not to be buried at all at these positions. The side chains of the ligands that bind to the T-cell receptor must not be buried into the MHC-binding cleft; hence they can be seen exposed to the solvent. This is true for the residues 4–7 of the active, high-affinity peptides.

Thus, the solvent-accessible surface correlates closely with the experimental observations and seems to be an excellent tool for careful and detailed interpretation of ligand–protein interaction.

7.4.4
Interaction Energies

The analysis of interaction energies has been broadly discussed and is still debated for instance in respect of hydrogen bond energies. Many authors consider that the calculation of interaction energies is useless because of the weakness of the potentials used for the calculation. The simple Coulomb equation is mostly taken for the electrostatics, hydrophobic interaction is neglected in most cases. Nevertheless, some calculations, if comparing several structures, at least provide an estimation, whether an energetically favored interaction occurs or not.

However, one should always bear in mind, that the computation of interaction energies sums up very large numbers to result tiny differences. This means, the calculation may be extremely sensitive to the choice of starting conditions, like geometries, or the quality of charge calculations.

In the present case, we have studied the energies of the protein–ligand interaction complex in terms of energy value per residue for charged residues, neutral residues, the whole peptide, and for the water molecules. The internal energy of each of these substructures has been computed every 20 ps throughout the molecular dynamics simulation.

The idea was that artefacts caused by strong interactions should be recognized by an increase of internal energy for the neutral amino acids, because those would have to compensate the structure artefacts. The analyses showed that all substructures had decreased their internal energies.

Most interesting was the finding that the energy of the bound peptides fluctuates quite widely but in general falls to a level that is more than 30% lower in energy than the starting structure which had been energy-minimized before docking. Again, this shows the performance of ligand optimization by molecular dynamics simulation in the presence of restraints from the binding site.

The complete situation has always been analyzed energetically. The vacuum, or solvent energy of the ligands may be very different from the minimum energy of the docked ligand. The docked conformation, however, is the only valid optimized structure which can be used for further drug design steps. The solvent conformations, and especially the vacuum conformations, are in most cases absolutely useless. There are of course exceptions, for instance for rigid molecules!

Also remarkable was the overall decrease in internal energy for the charged residues, which was computed to be about 60 kJ mol^{-1} per residue. The interactions of the bound peptide to neutral residues, or to itself, fluctuated and changed only slightly. This seems reasonable because during the docking process hydrophobic interactions were carefully optimized knowing the weakness of the potentials. Thus, no dramatic changes are to be expected during the molecular dynamics simulation of the interaction complex.

Although interaction energies might not be able to provide a quantification of the protein–ligand interactions—as had been previously hoped—we feel that they allow for an estimation of quality and provide a feeling for what happened during the molecular dynamics simulation.

There remains an open question whether more sophisticated calculations of interaction energies—for instance using "good" charge calculation that take into account the Poisson– Boltzmann equation—are able to improve the quality of interaction energy calculations. This is also true for quantum chemical interaction energy calculations. Although there are many counter-arguments, especially with respect to lacking performance of an ab initio basis set for non-covalent interactions, good results have been obtained even with very simple basis sets using quantum mechanical calculations. One such example is the prediction of ammonium partial structure interaction with an aromatic moiety by semiempirical methods as early as 1975 [12]. This interaction has been fully confirmed by the X-ray structure of acetylcholine esterase some 20 years later [13].

7.5
SAR of the Antigenic Peptides from Molecular Dynamics Simulations and Design of Non-natural Peptides as High-Affinity Ligands for a MHC I Protein

From the analyses described in the previous section, much information was provided for a design of non-natural ligands. Knowledge of the site, flexibility and side chain interaction of the binding pockets led to the idea to investigate whether all of them were used optimally by the native ligands. It evolved that the binding pocket D, responsible for interacting with side chains in position 3 of the peptidic ligands, provides much more space than the native ligands used to fill. This might provide the possibility of adding binding interactions by placing a larger side chain or substituent in this pocket. This would increase binding energy and hence lead to a ligand with high affinity.

Pocket D is hydrophobic in nature, lined by tyrosine, histidine, leucine, tyrosine and leucine. As seen in the crystal structure-based homology models, the ligand's side chains only interact with the upper rim of the hydrophobic pocket, taking into account only both tyrosines.

7.5.1
The Design of New Ligands

To rationalize the binding of a hydrophobic side chain to pocket D, we have computed the optimum interaction site. The computation was made for the isolated pocket using the program GRID (see section 4.6.2) and applying the methyl group as a probe. The resolution of the grid was 0.5 Å. Interaction of the methyl probe with the pocket walls was summed to result in a contour plot localizing negative binding energy. This indicates the approximate size of a putative ligand to fill the pocket completely and to interact with the pocket by gain of interaction energy (Fig. 7.13).

As we had hoped, the contour map extends much more deeply into the pocket, than the native ligand's side chain. Thus, it could accomodate even larger residues than phenylalanine for instance. Using molecular graphics, we were able to show that residues as bulky as naphthylalanine fitted nicely into the pocket. Furthermore, we predicted that this should be possible without distortion of the geometry of the peptide ligand's backbone.

The question whether the hypothezised interactions remain stable—as suggested by this static picture—was addressed by molecular dynamics of the solvated complexes. One of the bacterial peptides (Lys-Arg-Gly-Ile-Asp-Lys-Ala-Ala-Lys) was used as a template and position 3 substituted by apolar side chains of increasing size (Table 7.5).

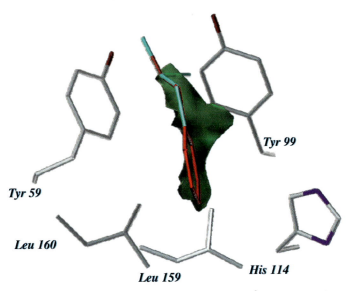

Tyr 99

Tyr 59

Leu 160

Leu 159

His 114

Fig. 7.13 Energy contours indicating at the 2.75 kcal mol^{-1} level the most favorable interactions between the free pocket D and a methyl probe. Complementary peptide side chains were fitted *a posteriori* to the energy contour map from the crystal structure of HLA-B27-bound nonapeptide (isoleucine, blue; homophenylalanine, red).

Tab. 7.5 Sequence of non-natural HLA-B27 ligands derived from two bacterial nonapeptides[c]

Peptide no.	Peptide sequence	Symbol
	Lys-Arg————Xaa————Ile-Asp-Lys-Ala-Ala-Lys	
1	Gly[a]	▲
2	Leu	▼
3	Hpa	◆
4	Ana	●
5	Bna	■
	Gln-Arg-Leu———— Spacer————Lys	
6	Lys-Glu-Ala-Ala-Glu[b]	
7	Gly-Gly-Gly-Gly-Gly	
8	Aba-Aba-Aba	
9	Aha-Aha	

[a] *Chlamydia trachomatis* groEl 117–125 [14].
[b] *Escherichia coli* **dnaK 220–228.**
[c] Hpa = Homophenylalanine; Ana = α-Naphthylalanine;
Bna = β-Naphthylalanine; Aba = 4-Aminobutyrie acid;
Aha = 6-Aminohexanoic acid.

Analysis techniques, described earlier, were then applied to determine, whether the constructs would be expected to be stable and worth synthesizing. Molecular dynamics simulation in water for 150 ps revealed for all cases that the new substituents in pocket D did not affect the 3D structure of the binding groove—at least, not to an extent that would destroy the complex. rms deviations from X-ray structure and from previous homology models, respectively, were reasonably low.

The analysis of the surface areas indicated the stability of the complexes with the new, non-natural side chains in position 3 (Fig. 7.14(a)). Buried surfaces (the oppo-

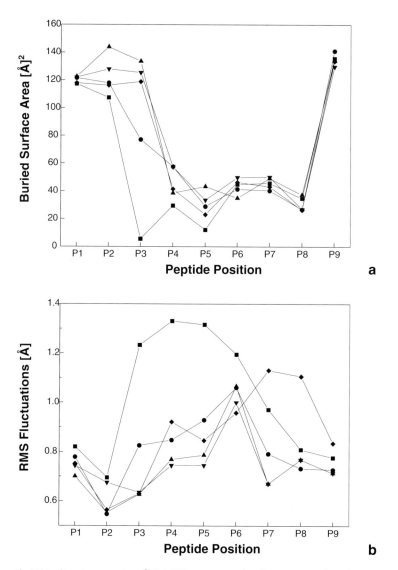

Fig. 7.14 Atomic properties of HLA-B27-bound nonapeptides 1–5 simulated by 150 ps molecular dynamics simulation on solvated MHC–peptide pairs. a) Buried surface areas of HLA-B27-bound peptides, calculated from relaxed time-averaged conformations. b) Rms atomic fluctuations averaged per peptide residue over all atoms and calculated from time-averaged conformations. Symbols used explained in Table 7.5.

site of solvent-accessible surface) of more than 100 Å^2 are found for the anchor positions 1, 2, and 9 corresponding to the binding pockets at the N- and C-termini. As expected the middle part of the ligand (residues 4–7) shows low buried surface areas. Insofar, the picture closely resembles the normal situation. A striking difference occurs at position 3 of the ligand, corresponding to pocket D. Correlated with the size of the residues the buried surface increases significantly up to 140 Å^2. By comparison the native glycine at position 3 is less than 10 Å^2. Thus, the analysis of the surface areas directly reflects the GRID calculations and the expectations arising from the calculations, that the non-natural side chains, as do the naphthyl derivatives, should be able to fill and stabilize binding at pocket D better than the native residue.

A very similar situation results from the analysis of the atomic fluctuations. As expected lowest fluctuations are found for anchor positions 1, 2, and 9. High mobility occurs at positions 4–7, representing the part of the peptide ligand pointing to the solvent or the T-cell receptor, respectively. At position 3, however, a clear discrimination is again possible between the native and the new, synthetic ligands. Atomic fluctuations at position 3 are strictly related to the size of the residue and hence to the number and strength of interactions to pocket D (Fig. 7.14(b)). Again, the most flexible ligand is the parent peptide having a glycine at position 3. In contrast, naphthylalanine residues restrain the whole peptide in its mobility by their tight binding to pocket D. Therefore, our prediction was to expect higher affinity for ligands bearing non-natural side chains in position 3. Those side chains should have aromatic/hydrophobic properties for optimal interaction with the residues that comprise the walls of the binding pocket in the MHC molecule.

7.5.2
Experimental Validation of the Designed Ligands

The non-natural ligands reported in Table 7.6 have been designed as described in detail in the previous section, in order to exhibit an enhanced affinity to HLA B*2705 through optimized hydrophobic interactions between the side chain of the peptide amino acid in position 3 (P$_3$) and the MHC protein binding pocket D. We use circular dichroism spectroscopy (CD) to monitor and quantify the differences between natural and non-natural ligands [14]. By stepwise heating up the MHC-pep-

Tab. 7.6 Sequence of the investigated peptide analog complexes.

Number	Peptide Sequence									T$_m$	
										B*2705wt	L156W
1	Lys	Arg	Gly	Ile	Asp	Lys	Ala	Ala	Lys	47.1	37.6
2	–	–	Phe	–	–	–	–	–	–	57.2	n.d.
3	–	–	Ana	–	–	–	–	–	–	62.3	35.2
4	–	–	Bna	–	–	–	–	–	–	51.9	41.5
5	–	–	Hpa	–	–	–	–	–	–	62.0	37.5

tide complex, thermal unfolding occurs as a function of stabilisation potency of the ligand. In other words, the better the ligand binds, the higher the structure stabilization that is to be expected. If the thermal unfolding follows a two-state equilibrium mechanism, the transition midpoint or melting temperature Tm is related to the free energy of folding by the Van't Hoff and Gibbs-Helmholtz equations. Therefore, differences in ΔG can be determined by measuring the thermal unfolding transition, e.g. as a change in circular dichroism.

If the same protein is tested with different ligands, differences in Tm should be related only to different intermolecular interactions energies; provided that all ligands tested share the same binding mode and do not alter the 3D structure of the protein. It has been shown previously that the thermal stability of class I MHC depends on the sequence of the bound peptide [15] and that Tm correlates with the equilibrium dissociation constant for a set of related peptides [16]. We investigated the thermal stability of MHC heterodimers in complex with single peptides, isolated by gel filtration chromatography. Assembly of MHC heterodimers occurs only upon peptide binding. As a prerequisite for the denaturation experiments, the reversibility of unfolding/folding was investigated. For this purpose, the CD spectrum of a HLA-peptide complex was measured consecutively at 25°C, after heating to 70°C and subsequent cooling to 25°C. The native fold could be easily identified by a strong negative dichroic signal at about 220 nm typical of an α-helix. The unfolding that occurred at high temperature, characterized by a spectrum typical of a disordered conformation, was found to be reversible, as shown by the comparison of pre- and post-heating spectra.

In accordance with our initial predictions, the designed non-natural peptides (Table 7.6) lead to a higher stability of the resulting complexes. All analogs of the chlamydial peptide 1, in complex with HLA-B*2705, show a higher Tm than the parent peptide. Not surprisingly, the substitution of Phe for Gly at P_3 also gives a substantial temperature shift towards a melting temperature of about 57°C, but the methylene homolog homophenylalanine 5, designed to fill the bottom of pocket D results in a further increase in stability of the same magnitude as that observed for the α-naphthylalanine analog 3. In contrast to α-naphthylalanine, a β-naphthyalanine (peptide 4) with a TM of about 52°C results in a slightly decreased melting temperature with respect to phenylalanine, though still constitutes a substantial stabilization. The slight decrease is obviously due to the different orientation of the naphthyl moiety and the lower steric complementarity for pocket D.

In order to validate our hypotheses for the binding events at pocket D in a further step, we mutated pocket D at its bottom exchanging a leucin side chain for a sterically much more demanding tryptophane side chain. This should result in most unfavorable conditions for binding large hydrophobic side chains in P_3 of the peptide, and we would hence expect a sharp decrease in stabilization with the mutant Leu 156 Trp. Either, if the non-natural analogs deeply bind to pocket D, their binding should be dramatically reduced by the mutation, or, if they bind non-specifically to the protein surface near the rim of the pocket, their affinity for the mutant should not be altered. The denaturation profiles of the mutant (Leu 156 Trp) in complex with our set of peptides are clearly in favor of a specific pocket-mediated binding mode. The mutant shows a reduced stability and nearly no differences in the melting temperatures of the natural and non-natural peptides. The large shift in Tm observed for the hydrophobic peptide variants and the absence of this shift in the pocket D mutant strongly suggests the validity of the predicted binding mode for the non-natural aromatic side chains. Since β-naphthylalanine differs from the other analogs in binding to the wild type protein, as reflected by a substantially smaller temperature shift with respect to the parent peptide 1, as well as in binding to the mutant, the P_3 side chain possibly exhibits a local alternative binding mode.

7.6
How Far Does the Model Hold? Studies on Fine Specificity of Antigene Binding to Other MHC Proteins and Mutants

Having established this first model of MHC protein-ligand interactions, it had been our next goal to apply this knowledge on the understanding of the molecular patterns of MHC-associated diseases. While reliable 3D models are now accessible and can be interpreted with respect to biological data [8,17,18]; it is still not proven that this molecular interpretation correlates with the resistance or susceptibility to human immunological diseases, associated with the expression of the respective alleles. Disorder such aus diabetes, rheumatoid arthritis or malaria [19], which are known to have a genetic prevalence, render this question highly important. This raises the question of how important the description of the whole ternary complex might be. All molecular models up to now have neglected the MHC-ligand-TCR interactions. In a series of studies we focused on disease-linked single residue mutations in the MHC-proteins, certain HLA subtypes showing altered ligand specificity and the influence of the T-cell receptor binding on the model. While our experiences in describing the fine tuning of the ligand interaction may be found in the relevant literature, the next section deals with the role of the T-cell receptor and consequences of its interaction.

7.7
The T-Cell Receptor Comes in

Whereas biological data are available to describe the mechanism of peptide processing and binding to MHC class I molecules, much less is known about the T-cell recognition of peptide MHC ligands. Numerous 3D structures of antibodies and of class I MHC are currently available, the structure of the ternary TCR-peptide-MHC complex, however, has long been obscure. Early comparison of TCR and antibody variable region sequences suggested a close similarity [20], confirmed several years later by first crystal structures of TCR variable domains. Within each TCR chain three complementary determining regions (CDRs) ensure diversity and were suggested to define the peptide specific MHC restricted combining site. The large sequence variation observed in TCR variable regions, as well as the relatively low pairwise amino acid identity to immunoglobulins (<25%), however, limited the use of Ig crystal structures as templates for building reliable three-dimensional models of TCRs. In the last five years, several X-ray structures of TCRs have been published, some of them in complex with MHC-peptide interaction complexes (e.g. see [21–24]). These data suggest that the TCR binds diagonally across the MHC binding groove [23]. Whether this crystallographic binding mode may be extended to all peptide MHC class I complexes is still a matter of debate. To generalize the experimental observations, more X-ray analyses would be required, preferably including mutants of the various components. Until such experiments have been conducted, it is reasonable to build TCRs by homology modeling, and in any event to be guided by such models in a rational approach to experimental work. Here, we have used existing crystal structures as templates to generate models of two new TCR-peptide-MHC structures representing two different T cells, both specific for the same peptide-MHC pair. These models have been examined in the light of a large body of experimental data obtained from the group of Buus and Engberg [25] containing an exhaustive set of 144 epitope analogs. This extensive feedback to experimental data supports the validity of the models and suggests a conserved recognition of MHC peptide ligands by T-cells.

The technical details of the experimental setup may be found in the literature [26]. Both commercially available software and programs developed in-house have been used for the interpretation of data and comparison of the two model complexes, respectively. Software packages comprise CARNAl within AMBER, PROCHECK for stereochemical reliability and Prosa II for correct folding [27,28]. Surface areas and non bonded interactions were examined using ACCES and CONTACT of the CCP4 software package [29].

Since the modeling of the MHC peptide complexes has been extensively described in the previous sections, we will focus here on the modeling of the ternary complex and the structural role of the TCR for only one of the types. Four X-ray templates (PDB: 1tcr, 2ckb, 1ao7, 1bd2) were chosen for multiple alignments of sequences to the 5H3TCR and to determine the best template for each chain. X-ray structures for unliganded Vα or Vβ chains, as well as V$\alpha\beta$ TCRs bound to antibodies were not taken into consideration for two main reasons:

1. The confirmation of the highly variable CDR loops is dependent on the presence and nature of the TCR ligand [24,30].
2. The side chains participating in Vα Vβ interface have different orientation in monomers and $\alpha\beta$ dimers [31].

Computation of sequence identities clearly favored one of the four candidates as best suited for building the Vα chain, whereas a second one turned out to be the appropriate one for building the Vβ chain. Docking in the subsequent step after homology modeling of the respective chain was performed in absence of the hypervariable loop CDR 3β. It should be remembered that the CDRs are the peptide-specific recognition site of the TCR, but in order to guarantee diversity they are also highly variable in sequence and highly flexible in conformation. They are thought to adopt the most appropriate conformational state for the specific peptide presented by the respective MHC allele. Residues 4–7 are bulging out of the MHC binding groove and are available for the TCR contact. This fact, however, makes it nearly impossible to predict the 3D structure of those binding loops of the TCRs CDRs with some reliability, as they are highly dependent on the peptide sequence presented. Hence we decided to dock the MHC-peptide complex to the 5H3-TCR homology model lacking the $\beta3$ loop first and finalized the homology model by adding the respective loop under spatial restriction of the docked ternary complex in recognition of the presented peptide. The β chain lacking the CDR3 loop was paired to the α chain according to the 1ao7 (PDB code) crystal structure.

Subsequently the V$\alpha\beta$ 5H3-TCR was energy-minimized and docked into the MHC-peptide complex according to the 1ao7 X-ray template. The rationale for choosing this particular template for docking of the ternary complex came from the suggestion that length and conformation of the (still missing) CDR 3 loop dictate the orientation of the MHC peptide ligand to be docked to the TCR. Its length, therefore, should be rather similar to the length of the template. This was true for 1ao7 but not for the others. Up to this point, the residues surrounding the missing CDR 3 β loop had been kept fixed.

Figure 7.15 depicts the presented peptide in red, bound to the MHC (green) and positions 4 and 7, respectively, interacting with the CDR 3 β loop (Vβ dark blue) of the T-cell receptor (TCR) at positions Val-97 and Gln-98.

Once the MHC peptide complex had been docked into the TCR, the missing CDR 3 β loop was built by a loop searching procedure.

Modeling the 5H3 TCR CDR 3 β loop by homology, the retrieved loop enabled the two polymorphic positions on the loop (Val-97 β and Gln-98 β) to be directed towards peptide residues P_4 and P_7 which exhibit complementary physicochemical properties and are of high relevance for TCR recognition.

It should be pointed out that the various procedures selected during the construction of the ternary complex lead to a 3D model of reasonable topology, *prior* to any energy refinement. The whole complex was finally solvated, energy-minimized and submitted to a restraint 500 ps MD simulation. The refined MD model differs from the starting homology model mainly in the orientation of the 6 hypervariable loops

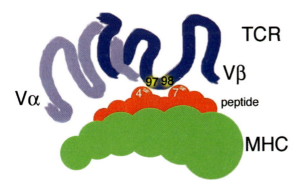

Fig. 7.15 The presented peptide (red) bound to the MHC (green) and its interaction in positions 4 and 7 with the CDR 3 β loop (Vβ dark blue) of the T-cell receptor (TCR) at positions Val-97 and Gln-98.

(α 1–3 and β 1–3) with respect to the MHC-peptide interaction partners, which was to be expected. However, the benefit of the MD refinement with regard to a model purely derived from homology concerns the interaction interface between the two proteins and the peptidic ligand. In the present case, MD allows for an optimization of intermolecular interactions, notably H-bonds. This may easily be explained by the choice of the CDR loop building procedure, based on TCR templates (PDB 1ao7 α chain and PDB 2 ckb β chain) that have a MHC peptide environment different from that of the target model. It may also be noted that the rms deviations from the starting structure, as well as positional atomic fluctuations were constant after 250 ps of MD simulation, so that time averaging of internal coordinates was indeed possible.

Subsequently, the validation of the model was achieved by testing the model for its ability to explain experimental observations. Without going step by step through the procedure of identifying every MHC-peptide-TCR contact and rationalize its importance, in summary we found only one disagreement between the model and the phenomenological binding data from mutagenesis experiments. The disagreement is related to the serine 3 of the antigenic peptide, the role of which in TCR recognition is obviously underestimated, at least according to the mutagenesis data.

What should be emphasized here, however, is once more the role of water. Atomic positional fluctuations of the 178 water molecules within the peptide-centered cap varied rather broadly from 0.4 to 7 Å, but none of them was further than 5 Å away from the ternary complex. This demonstrates the need for solvation of the complex which can easily be explained by the rather low complementarity between the surface of the TCR and that of the MHC peptide ligand leading to "holes" in the structure of the ternary complex that are filled by the solvent. Thus, bound water molecules contribute to enhancing the complementarity between the TCR and the MHC peptide ligand.

7.8
Some Concluding Remarks

Whereas everyone is aware of the fact that models do not represent reality and that their power is very often overestimated, the modeling procedure described here for the MHC case has two facets of general importance which are worthwhile in mentioning.

It has been shown that a careful and critical integration of theoretical and experimental data gives rise to new mechanistic findings even in retrospective models. The detailed interaction studies, not reported here in full length, led to the design of new biological experiments, emphasizing the importance of never anticipated long-range effects and the importance of completeness of models.

The second facet is that the model provided evidence for the sensitivity of putative MHC therapeutics against individual mutations. A majority of non-responders or even worse, "wrong-responders" are to be expected and hence might demand tailor-made ligands for each inividual patient. Those findings contributed to the rather critical design philosophy for MHC ligands and supported concerns voiced by clinicians. Thus, the modeling study had some impact as well on the general ideas of immunotherapy by blockade of MHC-T-cell communication.

References

[1] Nicklaus, M. C., Wang, S., Driscoll, J. S., and Milne, G. W. A. *Bioorg. Med. Chem.* **3**, 411–428 (1995).

[2] Rognan, D., Reddehase, M. J., Koszinowski, U. H., and Folkers, G. *Proteins* **13**, 70–85 (1992).

[3] Sufrin, J.,R., Dunn, D. A., and Marshall, G. R. *Mol. Pharmacol.* **19**, 307–313 (1981).

[4] Lambert, M. H., and Scheraga, H. A. *J. Comput. Chem.* **10**, 770–797 (1989).

[5] Fauchère, J. L., and Pliska, V. *Eur. J. Med. Chem.* **18**, 369–375 (1983).

[6] Rognan, D., Zimmermann, N., Jung, G., and Folkers, G. *Eur. J. Biochem.* **208**, 101–113 (1992).

[7] Madden, D. R., Gorga, J. C., Strominger, J. L., and Wiley, D.C. *Cell* **70**, 1035–1048 (1992).

[8] Rognan, D., Scappoza, L., Folkers, G., and Daser, A. *Biochemistry* **33**, 11476–11485 (1994).

[9] Rognan, D., Scappoza, L., Folkers, G., and Daser, A. *Proc. Natl. Acad. Sci. U.S.A.* **92**, 753–757 (1995).

[10] Honig, B., and Nicholls, A. *Science* **268**, 1144–1149 (1995).

[11] Connolly, M. L. *J. Appl. Crystallogr.* **16**, 548–558 (QCPE Programme No. 429).

[12] Höltje, H.-D., and Kier, L. B. *J. Pharm. Sci.* **64**, 418–423 (1975).

[13] Sussman, J. L., Harel, M., Frolow, F., Oefner, C., Goldman, A., Toker, L., and Silman, I. *Science* **253**, 872–875 (1991).

[14] Krebs, S., Folkers, G., and Rognan, D. *J. Peptide Sci.* **4**, 378–388 (1998). Rognan, D., Krebs, S., Kuonen, O., Lamas, J.R., Lopez de Castro, J.A., and Folkers, G. *J. Comput. Aided Mol. Design* **11**, 463–478 (1997).

[15] Bouvier, M. and Wiley D.C. *Science* **265**, 398–402 (1994).

[16] Morgan, C.S., Holton, J.M., Olafson, B.D., Bjorkum, P.J., and Mayo, S.L. *Protein Sci.* **6**, 1771–1773 (1997).

[17] Chelvanayagam, G. Jakobsen, I.B., Gao, X., and Eastel, S. *Protein Eng.* **9**, 1151–1164 (1996).

[18] Reveille, J.D., Ball, E.J., and Khan, M.A. *Curr. Opin. Rheumatol.* **13**, 265–272 (2001).

[19] Luyondyk, J., Olivas, O.R., Ginger, L.A., and Avery, A.C. *Infect. Immun.* **70**, 2941–2949 (2002).

[20] Chothia, C., Boswell, D.R., and Lesk, A.M. *EMBO J.* **7**, 3745–3755 (1988).

[21] Reinherz, E.L., Tan, K., Tang, L., Kon, P. Lin, J. Xiong, Y., Hussey, R.E., Smolyer, A., Have, B, Zhang, R., Joachimiak, A., Chang, H.C., Wagner, G., and Wang, J. *Science* **286**, 1913–192 (1999).

[22] Tissot, A.C., Ciatto, C., Mittl, P.R.E., Grütter, M.G., and Plückthun, A. *J. Mol. Biol.* **302**, 873–885 (2000).

[23] Ding, Y.H., Smith, K.J., Garboczi, P.N., Utz, U., Biddison, W.F., and Wiley, D. *Immunity* **8**, 403–411 (1998).

[24] Reiser, J.B., Gregoire, C., Darnault, C., Mosser, T., Guimezaues, A., Schmitt-Verhuls, A.M., Fontecilla-Camps, J.C., Mazza, G., Malissen, B., and Housset, D. *Immunity* **16**, 345–354 (2002).

[25] Stryhn, A., Andersen, P.S., Pedosen, L.O., Svejgaard, A., Helm, A., Thorpe, C.J., Fugger, L., Buus, S., and Engberg, J. *Proc. Natl. Acad. Sci. USA* **93**, 10338–10342 (1996).

[26] Rognan, D., Stryhn, A., Fugger, L., Lynbaek, S., Engberg, J. Ansoden, R.S., and Buus, S. *J. Comput. Aided Mol. Design* **14**, 53–69 (2002).

[27] Prosa II v-30: www.came.sbg.ac.at

[28] Sippl, M. *Proteins Struct. Funct. Gen.* **17**, 355–362 (1993).

[29] Collaborative computational project, No. 4: *Acta Crystallogr.* D50, 760 (1994).

[30] Housset, D., Mazza, G., Gregoire, C., Piras, C., Malissen, B., and Fontecilla-Camps, J.C. *EMBO J.* **16**, 4205–4216 (1997).

[31] Garcia, K.C., Degano, M., Pease, L.R., Huang, M., Peterson, P.A., Teylon, L., and Wilson, I.A. *Science* **274**, 209–219 (1996).

Index